WORLD BANK WORKING PAPER NO. 138

Accelerating Clean Energy Technology Research, Development, and Deployment

Lessons from Non-energy Sectors

Patrick Avato
Jonathan Coony

THE WORLD BANK
Washington, D.C.

World Bank Working Papers are published to communicate the results of the Bank's work to the development community with the least possible delay. The manuscript of this paper therefore has not been prepared in accordance with the procedures appropriate to formally-edited texts. Some sources cited in this paper may be informal documents that are not readily available.

ISBN-13: 978-0-8213-7481-8
eISBN: 978-0-8213-7482-5
ISSN: 1726-5878   DOI: 10.1596/978-0-8213-7481-8

Cover Photos: Bottom right: CFE's (Comisión Federal de Electricidad) La Venta II Wind Farm in Oaxaca, Mexico. Courtesy of Daniel Farchy, www.farchy.com. Remaining photos are the courtesy of the World Bank Group Photo Library.

Library of Congress Cataloging-in-Publication Data

Avato, Patrick, 1979-
    Accelerating clean energy technology research, development, and deployment :
lessons from non-energy sectors / Patrick Avato, Jonathan Coony.
        p. cm.—(World Bank working paper ; No. 138)
    Includes bibliographical references.
    ISBN 978-0-8213-7481-8—ISBN 978-0-8213-7482-5 (electronic)
    1. Renewable energy sources. I. Coony, Jonathan. II. Title.
    TJ808.V83 2008
    333.79′15—dc22

                                                                    2008013192

Contents

LIST OF FIGURES

Foreword

Climate change is one of the key challenges of this century. Specifically, balancing climate change mitigation and increased energy needs in developing countries poses a serious dilemma that can only be reconciled with new and improved clean energy technologies. However, to accelerate innovation in the energy sector, certain factors must be overcome, such as relatively low levels of research, development, and deployment (RD&D) funding and significant barriers to advancement. This paper addresses the necessary balance of climate change mitigation and energy needs while examining lessons learned from four case studies on new technology initiatives outside the energy sector.

In combating the impact of global climate change, the world faces unprecedented environmental, social, and economic challenges. As the Intergovernmental Panel on Climate Change's Fourth Assessment Report, the Stern Review, and other recent reports emphasize, the world risks devastating threats to our climate if no dramatic action is taken to reduce—*not just stabilize*—the levels of greenhouse gas (GHG) emissions. To compound the challenge, the need to reduce emissions comes at a time when the global economy is expanding and the worldwide demand for energy, infrastructure, and transportation is increasing rapidly.

Developing countries, as a group, have made impressive economic strides in recent years. However, energy use—the primary source of GHG emissions—is vital to their continued economic growth. At the same time, we must recognize that these countries are the least likely to be able to adapt to climate change.

Low-carbon energy technologies offer developing countries the best way to expand energy use to fuel their economies while simultaneously reducing global emissions. As new technologies become available, they can contribute to reconcile the choice between development *and* emissions reductions. Instead of following the same technological trajectories as industrialized countries, these countries can move directly to advanced clean technologies. Currently, however, most of the clean technologies available are too costly for widespread use.

To introduce new thinking in addressing these factors, this paper examines four cases from outside the energy sector where approaches to RD&D have been successful. These case studies highlight creative efforts in (i) international partnerships between public and private actors, (ii) information sharing and intellectual property rights, and (iii) novel financing schemes to generate valuable public goods.

As part of its commitment to fight poverty and promote development, the World Bank Group (WBG) has developed the Clean Energy for Development Investment Framework (CEIF) Action Plan. The CEIF outlines key activities the WBG is undertaking to mitigate GHG emissions and help client countries adapt to climate changes. Building on the successes and lessons of the CEIF, the WBG is now developing a comprehensive Strategic Framework for Climate Change (SFCC) to support developing countries' efforts to adapt to climate change and achieve low-carbon growth while reducing poverty. This paper contributes to an important part of WBG's climate change and energy work that focuses on analyzing the role of low-carbon energy technologies in climate change mitigation.

The four case studies presented in this paper are intended to stimulate thinking on novel approaches to clean energy technology development. They review approaches to innovation by the Consultative Group on International Agricultural Research, Advanced Market

Commitments for Vaccines, the Human Genome Project and the concept of Distributed Innovation. Although it is impossible to predict from which labs, universities, and businesses the critical technologies will emerge, it is clear that all countries must be more involved in advancing technologies and solutions. Many middle income countries are stepping up their technology development efforts and generating cutting edge clean energy technology innovations. It is critical to further expand these activities and also to involve low income countries from the onset to ensure that new technologies will be relevant to their needs and be ready for rapid deployment. This paper, along with an ongoing dialogue with stakeholders, can bring together the energy community to develop new approaches to clean energy and to begin meeting the challenges of global climate change.

Jamal Saghir
Director
Energy, Transport and Water
The World Bank

Acknowledgments

This paper was jointly funded by the World Bank Energy Unit (ETWEN) and the Energy Sector Management Assistance Program (ESMAP) and written by Patrick Avato (Energy Analyst) and Jonathan Coony (Senior Energy Specialist).

The authors wish to thank the peer reviewers of the report, Jeppe Bjerg (International Energy Agency), Corinne Figueredo (IFC), Lew Milford (Clean Energy Group), Alan Miller (IFC), Selçuk Özgediz (CGIAR), Alan Townsend (World Bank), and Alfred Watkins (World Bank).

Anil Cabraal, Ede Ijjasz-Vasquez, Jamal Saghir, and Gary Stuggins provided valuable strategic guidance and coordination. In addition, Jim Dooley, Richard Doornbosch, Michael Ehst, Hank Habicht, Greg Kats, Jeff Logan, Jessica Morey, Xiaoyu Shi, John Soderbaum, and Paul Runci provided important information and advice.

Marjorie Araya, Gabriela Chojkier, Thaisa Tiglao, and Janice Tuten assisted in processing and editing the final report.

Acronyms and Abbreviations

AMC	advanced market commitment
CCS	carbon capture and storage
CEIF	Clean Energy for Development Investment Framework
CFL	compact fluorescent light
CGIAR	Consultative on International Agricultural Research
CO_2	carbon dioxide
DOE	U.S. Department of Energy
EIA	Energy Information Agency
GHG	greenhouse gas
HGP	Human Genome Project
IEA	International Energy Agency
IARC	International Agricultural Research Center (CGIAR)
IFC	International Finance Corporation (of the WBG)
IGCC	integrated gasification combined cycle
IPCC	Intergovernmental Panel on Climate Change
IPO	initial public offering
IPR	intellectual property right
NIH	U.S. National Institutes of Health
NARS	National Agricultural Research Systems (CGIAR)
OECD	Organisation for Economic Co-operation and Development
ppm	parts per million
PV	photovoltaic
R&D	research and development
RD&D	research, development, and deployment
WBG	World Bank Group

Unless otherwise noted, all monetary denominations are U.S. dollars.

Executive Summary

*C*limate change is receiving considerable and increasing attention worldwide as one of the key challenges for the century ahead. In 2007 several new reports—including the Fourth Assessment Report of the Intergovernmental Panel on Climate Change (IPCC) and the Stern Review on the Economics of Climate Change—confirm and strengthen the evidence that climate change is indeed a real and serious environmental, social, and economic threat. These reports also underline a growing consensus that the efforts directed at mitigating climate change need a dramatic and timely increase to avoid potentially destructive and irreversible changes in the earth's climate.

To mitigate climate change, global greenhouse gas (GHG) emissions must be drastically reduced below business-as-usual levels and substantial reductions must begin in the next 10 to 20 years. Limiting human-induced climate change to two degrees Celsius above preindustrial era levels is viewed by many as the threshold before the risks of serious, irreversible impacts would rise exponentially. Because this goal appears increasingly difficult to achieve, much of the public discussion is now focused on how to limit temperature increases to a maximum of three degrees Celsius over preindustrial levels. Even to achieve this less stringent goal, the IPCC's Fourth Assessment Report estimates that the stock of CO_2 equivalent in the atmosphere would have to be limited to 550 ppm, requiring the annual flow of global GHG emissions to decrease: to a range of −30 percent to +5 percent from year 2000 levels by 2050. In contrast, business-as-usual projections show emissions increasing 25 percent to 90 percent already by 2030. Similarly, the Stern Review on the Economics of Climate Change estimates that global emissions must decrease 25 percent below current levels by 2050 and must peak in the next 10–20 years. To put this in context, from 1990 to 2005, global GHG emissions increased by 24 percent despite the fall in emissions in the former Soviet Union due to the economic collapse in the 1990s.

The need for massive emission reductions comes at a time when energy use—the primary source of GHG emissions—is expanding globally at unprecedented rates and is vital to the continued economic growth of client countries. The International Energy Agency (IEA) estimates that demand for primary energy will increase globally by 55 percent between 2005 and 2030. In developing countries, where economies and population are expected to grow fastest, primary energy demand is projected to grow by 74 percent during the same period. Fossil fuels are expected to remain the dominant source of primary energy, accounting for 85 percent of the overall increase in global demand.

This serious dilemma can only be reconciled with new and improved clean energy technologies that balance climate change mitigation and increased energy needs in developing countries. Better and broader use of existing clean energy technologies can play an important role in climate change mitigation. However, numerous sources identify new and improved clean energy technologies as essential to mitigate climate change while still allowing expanded energy supply as a key tool for further development. The IPCC Fourth Assessment, the Stern Review, and the IEA all state that sustainable levels of GHG emissions can only be achieved if new and improved technologies beyond those commercially available today are developed and deployed. Many of these technologies—in addition to reducing emissions—will improve energy access and energy systems reliability and reduce the impact of high and volatile fossil fuel prices.

While there are many promising clean energy technologies, most are currently too costly, lack the technical reliability needed for widespread deployment, or both. Energy technologies, both currently in use and under development, have the potential to reduce carbon emissions substantially. Such options include renewable energies, carbon capture and storage, more efficient power generation from fossil fuels, nuclear power, and improved efficiency of end-use technologies, industry, and transport. Currently, comparatively high costs and insufficient operating experiences very often hamper deployment on the scale needed for climate change mitigation and to meet rising global demand for primary energy.

The research, development, and deployment (RD&D) activities needed to commercialize these clean energy technologies have—after a period of significantly reduced activity—increased substantially over the last two to three years. From the mid-1980s to the early 2000s, energy research and development (R&D) spending was well below historic highs. By 2003, public energy R&D spending in the OECD had fallen to 60 percent from its peak in 1980, and private sector spending had fallen from $8.5 billion in the late 1980s to $4.5 billion in 2003. While absolute investments in energy innovation continue to lag behind historical levels, the trend appears to be reversing as concerns about climate change, energy security and high oil prices are prompting intensified private and public R&D activities.

However, these renewed efforts will face significant barriers that impact the ability to develop and deploy promising clean energy options. The following factors discourage investment in clean energy RD&D and reduce its effectiveness:

- *Uncertain future value of CO_2 emissions abatement.* Frameworks for CO_2 valuation continue to evolve but political and market risks, as well as post-2012 concerns, undermine the long-term planning needed to justify expensive RD&D.
- *Mitigation of climate change is a global public good.* Provision of a public good is hampered by free-riding across space—countries that free-ride on the mitigation efforts of others—and across time—actors that avoid the costs of mitigation now because the benefits of timely mitigation will be reaped by future generations.
- *The "Valley of Death,"* which occurs when promising technologies languish between public and private sector RD&D efforts.
- *Intellectual property rights (IPRs).* The large RD&D investments needed for technical advances in clean energy will be undermined by uncertain global IPR protection. At the same time, IPRs also hamper the deployment of technologies once commercialized.
- *Challenges developing and transferring technology to developing countries.* Developing countries—a substantial source of incremental emission growth—will need OECD resources and expertise to deploy the needed clean energy technologies.
- *Subsidies for conventional energy products.* Subsidies at both the retail and production levels reduce to below-cost the price with which new energy technologies must compete. Moreover, deployment of clean energy technologies is often hampered by trade barriers.

These factors must be overcome to accelerate innovation in the energy sector.

To overcome these barriers and introduce new technologies that deliver the massive scope of emissions reductions in the urgent time frame required, new and creative approaches to energy RD&D will be needed. Previous energy RD&D has not created the scale of technology transformation now required despite RD&D spending levels substantially higher than today. While increased spending will certainly be required, creative approaches and novel paradigms beyond traditional RD&D vehicles will be necessary to accelerate energy technology innovation on the scale and in the time frame required. The time is right for introducing new RD&D approaches to inform and influence many of the new initiatives now being launched.

To introduce new thinking in addressing these concerns, this paper examines four cases from outside the energy sector where creative approaches to RD&D have successfully overcome similar barriers:

- *The Consultative Group on International Agricultural Research (CGIAR),* which played a key role in the "green revolution" and continues to support agricultural research for developing country needs;
- *Advanced Market Commitments (AMCs),* which pool OECD donor funds to provide incentives to the pharmaceutical industry to develop vaccines for tropical diseases;
- *The Human Genome Project (HGP)* which, through public and private participation, fully decoded the entire human genome two years ahead of schedule; and
- *Distributed Innovation,* where large numbers of dispersed people and companies contribute to innovation cooperatively using nontraditional forms of intellectual property rights.

Lessons learned from these case studies provide important insights that can be applied to accelerate the commercialization of clean energy technologies.

- *Serving a public good.* The case studies show how to develop technologies that serve a global public good through mission-oriented research and creative agreements to coordinate contributions of multiple governments while enticing private sector participation at an early stage.
- *Facilitating innovative research partnerships.* The case studies show the importance of innovative partnerships (such as public-private, international, and North-South for technology transfer) and provide model structures that can encourage them.
- *Overcoming the "valley of death."* The case studies suggest several ways to address the "valley of death" problem, for example, via donor subsidies that are only paid if industry develops the desired, not-yet-available technology.
- *Technology Transfer: North-South and South-South.* The case studies show ways to marshal OECD resources and expertise to serve the technology needs of developing countries, as well as vehicles for South-South information sharing.
- *Potential World Bank Group contributions to technology development.* Two of the case studies—CGIAR and AMC—show how WBG strengths can be used to catalyze the commercialization of technologies that serve client country needs and are not otherwise provided by public and private actors.

These lessons learned can strengthen the development of new technology initiatives, which will address the needed balance between climate mitigation and the growing energy demands of the developing world.

CHAPTER 1

Introduction

Climate change is receiving considerable and increasing attention worldwide as one of the key challenges for the century ahead. In 2007 several reports have been published that confirm and strengthen the evidence that climate change is indeed a real and serious environmental, social, and economic threat. These reports, including the Fourth Assessment Reports of the Intergovernmental Panel on Climate Change (IPCC 2007a–c) and the Stern Review on the Economics of Climate Change (HM Treasury 2006), underline a growing consensus that the efforts directed at mitigating climate change need a dramatic and timely increase to alleviate potentially destructive and irreversible changes in the earth's climate.

The World Bank Group's Clean Energy for Development Investment Framework Action Plan (World Bank 2007a) has outlined some of the key activities it intends to undertake in the area of mitigating greenhouse gas emissions and helping client countries adapt to changes in climate.

One of these activities focuses on an analysis of the role of low-carbon energy technologies in climate change mitigation. This report provides an initial analysis of this issue. Chapter 2 describes the urgency of developing new low-carbon energy technologies based on a review of some of the most authoritative recent reports on climate change. Strong evidence demonstrates the need for new and improved energy technologies, but—as is described in Chapter 3—current research, development, and deployment (RD&D) efforts worldwide appear too limited and slow-paced to generate new energy technologies rapidly enough to respond to the climate change crisis. Moreover, significant barriers are limiting incentives to invest in energy RD&D and may reduce the effectiveness of such investments. These barriers are discussed in Chapter 4. In light of these barriers and the very limited success of past attempts to overcome them, Chapter 5 then analyzes four case studies where related barriers have been successfully overcome and public goods have been generated in non-energy sectors. These case studies are purposefully drawn from non-energy sectors to introduce new thinking to the energy sector and develop lessons learned to inform the development of novel and creative energy innovation vehicles. Chapter 6 draws lessons from these case studies that speak to creative ways to approach RD&D. The final chapter summarizes findings and makes suggestion for follow-on work.

CHAPTER 2

Climate Change and the Need for New Clean Energy Technologies

The Growing Global Concern about the Threat of Climate Change

The surge has been remarkable in the call for action to combat climate during 2007. Most notable has been the conclusion of the Fourth Assessment of the Intergovernmental Panel on Climate Change (IPCC 2007b) that warming of the climate system is "unequivocal" and very likely due to the observed increase in anthropogenic greenhouse gas (GHG) emissions.

The evidence is now robust that the stock of GHG emissions in the world's atmosphere is directly related to the flow of human-induced GHG emissions. These are predominantly caused by the energy supply sector; transport; industry; and land use, land use change, and forestry (LULUCF). The energy sector is one of the largest emitters and thus absolutely essential to incorporate in any mitigation strategy. In fact, energy supply has been the fastest growing source of emissions from 1970 to 2004 (+140 percent) and the great majority of projections (for example, the IEA projections in figure 1) expect these emissions to grow even faster in the future. These emissions are generated by the combustion of carbon-intensive fossil fuels such as coal and oil and—to a significantly smaller extent—natural gas. OECD countries are currently the major source of GHG emissions and they have contributed the majority of the stock of GHG currently in the earth's atmosphere. However, due to their rapidly growing economies, size, and cheap and abundantly available fossil fuel resources such as coal, energy-related GHG emissions from developing countries are expected to increase dramatically. The IEA estimates that demand for primary energy will increase globally by 55 percent between 2005 and 2030 and by 74 percent in developing countries (IEA 2007c).

A consensus is forming in the world that global warming should be limited to a 2–3°C rise in temperature from preindustrial era equilibrium to avert the most serious impacts of climate change. The models that have been used to analyze the impact of this target indicate that this limit requires the stock of CO_2 equivalent atmospheric concentrations not to

3

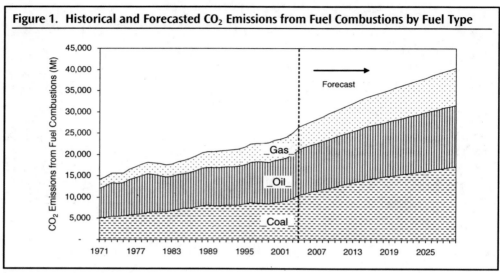

Figure 1.   Historical and Forecasted CO$_2$ Emissions from Fuel Combustions by Fuel Type

Source: Adapted from IEA 2006b and 2007a.

exceed 550 ppm. To accomplish this target requires that GHG emissions peak between 2010 and 2030 and that by 2050 global CO$_2$ emissions be dramatically lower than under business-as-usual projections. IPCC estimates that a stabilization at 550 ppm will require global GHG emissions to peak between 2010 and 2030; by 2050 emissions will then have to decrease: to a range of −30 percent to +5 percent relative to 2000 levels. As a comparison, IPCC's business-as-usual scenarios estimate that already in 2030 GHG emissions will increase to levels 25–90 percent higher than in 2000. Similarly, the Stern Review estimates that global GHG emissions will have to peak in the next 10–20 years and then decrease to 25 percent below the current level by 2050. To put this challenge in context, from 1990 (when the IPCC First Assessment was delivered) until 2005, global GHG emissions increased by 24 percent.

Based on this growing evidence, the IPCC, the Stern Review on the Economics of Climate Change (HM Treasury 2006), and the International Energy Agency (IEA 2006a) conclude that to mitigate the severity of climate change impacts on developing and developed countries alike, GHG emissions must be dramatically reduced. To give an idea of the magnitude of this needed reversal, Figures 1 and 2 show the substantial historical and projected CO$_2$ emissions from fuel combustion broken down by fuel type and region.

Reversing this trend will pose a huge challenge for all parts of the world economy, specially the energy sector, which will have to shift massively to low-carbon technologies within the next 10-20 years.

Clean Energy Technology Options

Major changes from business as usual are needed to shift the energy sector onto a sustainable track. All major reports on climate change confirm that such a shift will require some mix of the following clean energy technologies:

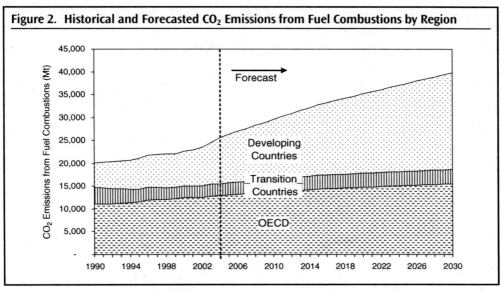

Figure 2. Historical and Forecasted CO$_2$ Emissions from Fuel Combustions by Region

Source: Adapted from IEA 2006b and 2007a.

- increased energy efficiency in power supply, demand, and transport
- renewable energy—including wind, hydro, solar, and geothermal power and biofuels
- nuclear energy
- fuel switching to less carbon-intensive fuels (for example, from coal to natural gas)
- carbon capture and storage (CCS).

In addition to these "hardware" technologies, "software" technologies including innovations in information technology, management, and planning (such as urban planning) can also play a critical role in mitigating climate change.

Clean energy technologies differ substantially in their aggregate potential to reduce emissions (a function of the absolute availability of the resource and relative costs) and in terms of the stage of their development (such as whether the technologies have already been commercially proven or not). Figure 3 illustrates the most widely discussed forms of clean energy supply along these two dimensions, indicating the stark differences among the technologies. Appendix A provides a brief description of the innovation chain depicted along the x-axis of the figure. Appendix B provides descriptions of the main assumptions behind the emission reduction potential of each technology as estimated by IEA and of the most promising clean energy technologies under development.

In addition to stage of development and GHG emissions mitigation potential, clean energy technologies differ along other dimensions with strong implications for the design of policy and investment instruments aimed at climate change mitigation, including the following:

- *Diversity within each technology category.* Many of the technologies commonly described as clean energy technologies actually consist of a wide array of different technologies or technology applications, each at a different stage of development

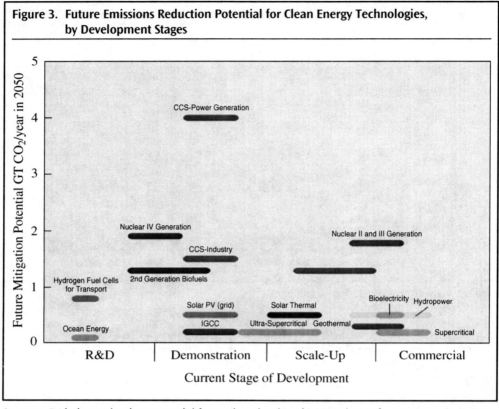

Figure 3. **Future Emissions Reduction Potential for Clean Energy Technologies, by Development Stages**

Sources: Emission reduction potential for each technology in 2050 drawn from IEA 2006a. Stages of technical development are World Bank staff estimates.

Note: The potential reduction refers only to *increased* emission reductions in year 2050 in Gt/year compared with the base case scenario that would result from accelerated technology development and deployment, and a package of policies that lead to adoption of technologies that reduce CO_2 emission at $25/ton.

Solar: The potential shown for each solar technology represents the *combined* emission reduction potential development for both PV and solar thermal technologies. It has been split into two bubbles to show different stages of development for each technology. Supercritical and Ultra-supercritical: The potential for supercritical and ultra-supercritical represents the *combined* emission reduction potential from both technologies. It has been split into two bubbles to show different stages of development for each technology.

and with distinct characteristics. For example, hydropower ranges from large hydro-electric dams to smaller run-of-river facilities to microhydro plants, each with distinctly different technical, economic, social, and environmental characteristics. Similarly, solar photovoltaic (PV) faces considerably different challenges when applied in off-grid applications or connected to a grid.

■ *Structure of the supply sector.* The supply industries that manufacture each of these technologies differ substantially. For some, only a handful of large, mostly multina-tional companies are capable of RD&D or actual manufacturing, for example, with nuclear power or carbon capture and storage (CCS). For others, numerous players of various sizes scattered around the globe can and do play a role in the technology's technical advancement and manufacture, for example, with many end-use technologies.

- *Structure of demand sector.* The consumers both of the technologies themselves and the products generated by those technologies differ substantially. For large power plants, such as ultra-supercritical coal stations, a finite amount of large power generators (primarily utilities) would even consider purchasing one. This effective global oligopsony can strongly influence, and in fact participate in, the direction any development effort may take. Retail products such as compact fluorescent lights (CFLs), however, are sold to millions of consumers through extensive distribution networks and various intermediaries.

- *Capital intensity of R&D.* The amount of money needed to fund R&D for each technology differs substantially. For example, design and construction of a single CCS demonstration plant costs hundreds of millions of dollars. Other technologies, such as most energy-efficient end-use products, can gain incremental yet important advances from the work of small labs with relatively small sums of money.

- *Required adaptation to local conditions.* Some technologies will have a global reach while others can only be used under certain conditions or will need to be adapted substantially when being transferred from one region to another. Differences in climatic and natural resource conditions play an important role for many forms of renewable energy, demand density affects the design of electricity systems, and differences in lifestyles and living conditions affect the types of appropriate end-use technologies.

- *Production of intermediate or end-user product.* In some cases the products from these technologies are intermediate forms of energy that are then used by the final consumer, for example, electricity generators. In other cases the products from these technologies provide the final useful energy form for the final consumers, for example, most end-use technologies such as light bulbs and electric motors.

- *Other nontechnical factors.* In addition to the technical dimensions described above, technologies differ substantially in how they are affected by nontechnical and institutional factors. Nuclear power and hydropower, for example, can have negative nonCO_2 environmental and social costs and risks. Energy-efficient technologies face a host of barriers to deployment such as lack of information, the landlord-tenant (principal-agent) problem, and an inability or unwillingness by consumers to properly consider lifetime costs when making equipment purchases.

The Need for New and Improved Clean Energy Technologies

Despite these important differences, clean energy technologies have in common that the more widespread deployment of most of these technologies is constrained by comparatively high cost and reliability constraints in many applications. For example, new coal technologies—such as integrated gasification combined cycle (IGCC), supercritical, and ultra-supercritical—significantly reduce emissions compared with baseline coal technologies but in most cases they are considerably more expensive. Renewable energies such as solar power and wind generate no emissions during operation but suffer from the intermittency of wind and sunlight and are generally higher-cost than more polluting options for electricity grids.

Consequently, while the more widespread adoption of existing technologies can have significant mitigation potential and should definitely be pursued, the IPCC, the Stern Review, and the IEA all conclude that a truly sustainable energy future can only be achieved if new and improved clean energy technologies beyond those commercially available today are developed. In fact, each of these reports examines the mitigation effects of policies to control emissions, including increased deployment of commercial or near-commercial clean energy technologies, and concludes that while helpful, such efforts will be insufficient unless combined with an accelerated commercialization of new and improved clean energy technologies (see Appendix C). Moreover, these studies emphasize the urgency with which new and improved clean energy technologies have to be developed and deployed. Among other factors, a large increase in funding for energy research, development, and deployment (RD&D) is needed.

CHAPTER 3

Trends in Energy Research and Development Spending

This chapter provides an overview of energy R&D trends. See Appendix D for more detailed data and graphs.

A Period of Reduced Energy R&D Spending from Mid-1980s to Early 2000s

Energy R&D spending by governments and the private sector has been well below historic highs consistently since the late 1980s. The late 1970s and early 1980s saw high levels of global energy R&D spending in response to concerns about high oil prices and energy security in the wake of the oil shocks. Thereafter, a period of low oil prices, reduced concerns about energy security, and energy sector market reform combined to significantly reduce interest in new energy technologies from the mid-1980s to the early 2000s. By 2003, energy R&D spending by OECD governments—the major funders of energy R&D—had fallen to 60 percent of its peak in 1980 (in real terms, see figure 4).

At the same time, the private sector also reduced their energy R&D activity, which fell globally from about $8.5 billion at the end of the 1980s to about $4.5 billion in 2003 (both figures in 2004 dollars). Measured in terms of R&D expenditures as a share of turnover, and considering the capital intensity of the sector, energy is one of the world's least innovative industries. Throughout the 1990s, energy R&D intensity fell and in 2002 stood at slightly more than 0.15 percent. This compares with an average R&D intensity of 2.6 percent for the manufacturing sector as a whole and more than 10 percent in the high-tech sector (Doornbosch and Upton 2006).

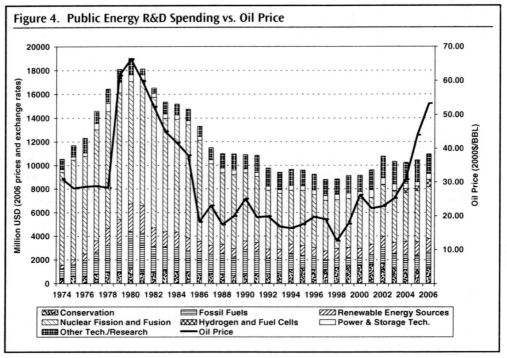

Figure 4. Public Energy R&D Spending vs. Oil Price

Source: Adapted from Doornbosch and Upton 2006.

Renewed Public and Private RD&D Activity in Recent Years

However, this trend of low energy RD&D appears to be reversing as concerns about energy security, climate change, and high oil prices are leading to renewed public and private activity over the last few years. UNEP's 2007 report *Global Trends in Sustainable Energy Investment 2007* estimates that R&D spending by renewable energy and energy efficiency by governments and corporations rose from $13 billion in 2005 to $16.3 billion in 2006.

Government funding for energy RD&D has increased in most OECD countries. Of the 17 (out of 26) IEA countries for which reliable country-by-country public energy RD&D data are available for both 2004 and 2005, 13 show an increase in spending (in real terms) from one year to the next, while only 4 show a decrease. While the rise in cumulative spending for all 17 countries (in real terms) from 2004 to 2005 is only 2.3 percent, anecdotal evidence suggests continued increases in national energy RD&D funding in the major OECD countries through 2006 and 2007.[1] More recent information on energy RD&D activities from the two governments that dominate energy spending in this field, Japan and the United States, confirms the increase in public efforts (see Appendix B).

In addition to these renewed national efforts, two governments have recently proposed the development of facilities aimed at the development and deployment of clean energy technologies in developing countries. In September 2007, the U.S. government proposed

---

1. RD&D data for these years is not yet available in a standardized and form but several countries for which government data have been released indicate a continuation of this trend.

the creation of a new international clean technology fund targeted to help developing nations harness the power of clean energy technologies. This fund would help finance clean energy projects in the developing world and would be open to contributions from a variety of donors. The United Kingdom has announced an International Environmental Transformation Fund endowed with £800 million to help developing countries respond to climate change and protect their environments. Similarly, the U.S. government planned to begin exploratory discussions with participating countries.[2]

There have also been substantial increases in other means of government support for clean energy technologies beyond those associated with traditional R&D programs. Primary among these are support systems for renewable energy that can take the form of direct subsidies, feed-in tariffs, renewable portfolio standards, tax rebates, and biofuel blending mandates. These support policies may at least in part compensate for the relatively low R&D figures described above and have been increasing almost without exception in OECD countries in the last two to three years. Such deployment incentives amount to an estimated $33 billion globally (HM Treasury 2006). Such support leads to fuller deployment of technologies not yet commercially competitive, such as wind power, and in this way brings manufacturing and operating experience that leads to more reliable, lower-cost technology. The increase in such support systems thus acts as a *de facto* technology acceleration programs, although it must be noted that—in contrast to RD&D programs—they apply only to technologies with near-term commercial viability and reliability.

Private sector interest and investment in clean or alternative energy sources has also increased in the past few years. The total funds raised through IPOs of renewable energy and energy efficiency companies reached $8.4 billion in 2006, up from $3.5 billion in 2005 and only $0.4 billion in 2004. In addition, since August 2004, 19 different index and mutual funds devoted to clean or alternative energy have been launched in the United Kingdom or the United States. These funds have averaged an annualized return of 34 percent from their start dates through early October 2007. While these figures do not explicitly split out the RD&D components of these investments, they do reflect very well the trends and interests in the sector. Similarly, global venture capital investments in clean technology companies surged to $1.1 billion in the first six months of 2007 and the investments are estimated to increase by more than 35 percent in 2007 as compared with 2006. Since 2001, clean technology's global share of overall venture capital investments has more than doubled.[3] According to the Cleantech Venture Network, a group that tracks capital flows to clean technology companies, investments in this field have increased 33 percent from 2004 to 2005 and a further 80 percent from 2005 to 2006.

The Increasing Role of Rapidly Growing Client Countries in Energy RD&D

With rapid rates of economic growth, increasing technical sophistication, and growing investment needs in energy infrastructure, various developing countries are increasing investments in domestic energy RD&D. In addition to climate change, such policies and investments are often developed to address concerns about the stability of energy systems through a

---

2. White House press release, September 27, 2007.
3. *CNN Money*. http://money.cnn.com/news/newsfeeds/articles/prnewswire/NEW04626092007-1.htm.

diversification of the energy portfolio. Moreover, R&D often focuses on technologies targeted to specific local conditions such as distributed low-load demand, limited maintenance capability, or specific locally available resources.

While data on these RD&D activities are even more difficult to find that for IEA countries, anecdotal evidence strongly indicates the growing impact of energy policies and R&D activities in developing countries on global clean energy markets. For example, China has developed a national climate change plan that calls for increasing the overall contribution of renewable energy to energy supply from about 7 percent today to 10 percent by 2010 and 16 percent by 2020. Moreover, during the past two decades billions of dollars in foreign technology purchases have helped to seed Chinese domestic R&D programs in the energy sector (Karplus 2007). Similarly, indigenous energy-related RD&D in India has been increasing substantially in the past years and the Energy ministry has recommended that the National Planning Commission include large funding for renewable energy in the 11th National Five Year Plan (2007–12). Proposed activities include increasing the share of hydro, nuclear, and renewable sources in the energy mix; developing more efficient coal technologies; improving energy efficiency in industry and transport; and exploring hydrogen (production, storage, and end-use) technologies.[4] Also Brazil has expanded its long-standing biofuels program with other energy-related support policies. In 2004 the federal government enacted the Alternative Sources of Energy Incentive Programme (PROINFA) Law, which aims to install additional capacity of 3,300 megawatts in wind, small hydro, and biomass energy.

The Limits of Renewed Energy RD&D Activity

However, despite this apparent worldwide rise in public and private RD&D investments, it is not obvious that these efforts will produce new clean energy technologies in the sufficiently short time frame that the urgency of climate change mandates. In fact, both the IPCC and the Stern Review strongly argue for massive increases in RD&D funding by governments and the private sector—to levels far higher than the latest data suggest. Similar calls have recently also been reiterated by the World Energy Council (2007) and the Inter Academy Council (2007). The Stern Review also acknowledges the barriers to innovation that must be addressed beyond mere increases in funding and suggests that funding be complemented with policies that tackle the barriers to technology innovation particular to the energy sector. In addition, experience with government-funded RD&D programs suggests that—while essential—large amounts of government funding are not enough to accelerate technological development so that widespread deployment allows emissions to begin substantial reductions in 10 to 20 years. Even at the apex of global government energy RD&D funding in the early 1980s the results of government energy programs did not deliver technologies that led to the kind of wholesale transition in the energy sector now required and emissions continued to rise strongly. Clearly, novel approaches are needed to complement traditional energy RD&D programs, improve their effectiveness, and successfully leverage and facilitate private sector activity in this area.

---

4. http://planningcommission.nic.in/aboutus/committee/wrkgrp11/wg11_rdenrgy.pdf.

CHAPTER 4

Barriers to the Development and Deployment of Clean Energy Technologies

Numerous factors are responsible for the limited success of traditional RD&D programs in the energy sector and for discouraging private sector investments. This chapter provides an overview of the most important macro-barriers that need to be overcome to increase the effectiveness of energy RD&D. It should be noted, however, that in addition to the barriers discussed here, numerous other context-specific micro-level barriers hamper the development, transfer, and diffusion of new technologies, especially in developing countries (see for example, World Bank 2008).

Negative Externality of Carbon Emissions Is Difficult to Valuate

There is no reliable valuation for the primary product that clean energy products deliver: reduced emissions. Consequently, the private sector has little incentive to develop cleaner technologies that reduce emissions. Nearly all aspects of energy production, transformation, and use result in CO_2 emissions that accumulate in the atmosphere. This negative externality is not valued and therefore not factored into investment decisions by energy providers, particularly by the private sector. As a result, energy-related decisions are made without reflecting their full costs and investments in clean energy technologies are lower. A patchwork of policies is slowly evolving; the current system is not easily predictable over the medium- to long-term and does not include the majority of global emissions. The magnitude and scope of any future carbon valuation is still subject to substantial political risk. Consequently, the future value of GHG emissions is discounted significantly in investment decisions and does not yet provide sufficient incentives for RD&D investments.

Climate Change Mitigation Is a Global Public Good

Climate change mitigation is a classic example of global public good: emission reduction from a single party benefits that party and everyone else. Although the costs of abating GHG emissions are borne by specific entities, no country or private actor can be excluded from the benefits. Emission reduction efforts invite free-riding behavior across space (by some countries free-riding on the efforts of others) and across time (by countries and private actors avoiding the costs of mitigation now because the benefits of timely mitigation will be reaped by future generations). This free-riding extends to energy technology development because the benefits accrued to any entity represents only a small share of the global benefits. Consequently, the private sector and governments will tend to underprovide for clean energy RD&D.

The "Valley of Death" between Public- and Private-Sector Development

The public and private sectors both play important roles in technology commercialization. The public sector normally begins the process with basic scientific research, often without specific end-products in mind. Much of this research is lab-based and involves high risks with potentially high return with great uncertainty of the impacts. As technical challenges are overcome and product ideas appear with profit potential, the private sector becomes increasingly involved. The risk is still high—as are the potential returns—at this stage but the products under development are now much closer to commercial application.

   The "valley of death" refers to the period in product development *between* public and private sector involvement (see Figure 5). The major technical problems have been solved in the lab and one or more potentially profitable products have been identified. At this point, public-sector participation usually declines as the government backs away from "picking

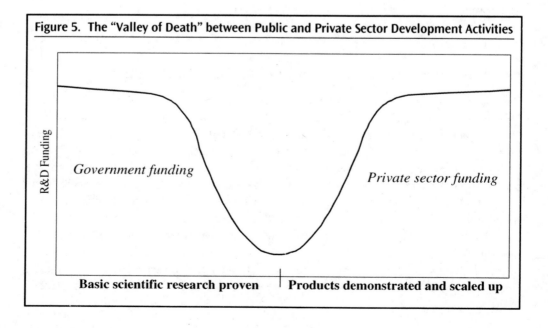

Figure 5. The "Valley of Death" between Public and Private Sector Development Activities

winners." Furthermore, governments do not want to subsidize private industry or distort the market because investors stand to profit handsomely from the ultimately commercial products. However, the private sector often still sees too much risk to get fully involved and to continue the product development process on its own (see "mountain of death" in the following section); promising technologies do not progress to the demonstration and scale-up stages needed to achieve full commercialization. Although many technologies do eventually make it through this period—either through additional public, private, or combined efforts—this gap seriously delays commercialization and prohibits the rapid deployment of clean energy options in the urgent time frame needed.

The "Mountain of Death" of Technology Costs

Technological innovation requires progression along the learning curve. Initial steps focus on component parts of a larger desired technology. Because these components represent only a portion of the total system, funding to develop and test them will be small compared with the full cost of the entire system. After researchers achieve some level of success with individual components, the next step is to integrate the components into the full system. The first full integration represents the highest per-unit cost that the developers will likely face. As more is learned about both the system as a whole and the individual components, and as economies-of-scale are achieved in the manufacturing costs, per-unit costs will fall. Eventually the technology reaches maturation, at which point the per-unit costs will be sufficiently low and technical reliability will be sufficiently high to warrant continued manufacturing of a commercial product.

This process of rising and then falling per-unit costs is referred to as the "mountain of death" for new technology innovation (see figure 6). It can deter R&D by requiring

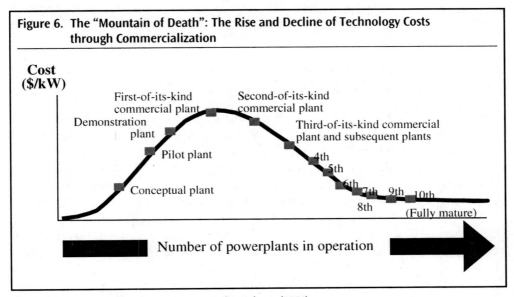

Figure 6. The "Mountain of Death": The Rise and Decline of Technology Costs through Commercialization

Source: Courtesy of Electric Power Research Institute (EPRI).

substantial upfront costs to develop and build products that, for a while at least, are not commercially viable. The "mountain of death" is a key reason that private companies are reluctant to invest in pilot and demonstration plants, and thus contributes significantly to the "valley of death" phenomenon.

Technology Needs of Developing Countries Are Not Adequately Served

Because large shares of emissions are projected to come both from developed and developing countries, emission reductions must be achieved in all regions of the world. Developing countries' share of global emissions is projected to increase from 39 percent in 2004 to 52 percent in 2030 (IEA 2006b). However, the incentives are low to develop clean energy technologies to serve the specific needs of these countries. While many energy technologies can be applied in all countries regardless of their conditions (such as a behind-the-fence CCGT power plant), other technologies must be adapted to local circumstances. Technologies to be adapted include end-use technologies suited to local consumption patterns, technologies to operate in the absence of a strong operations and maintenance support network, technologies to serve distributed generation or low-energy density-demand areas, and resource mapping for renewable energy sources. The vast majority of resources and expertise for technical innovation are in the OECD countries, which are rarely motivated to develop products suited solely for the poorer countries. OECD governments will naturally promote research that serves their own citizens and economies. Similarly, private industry often regards markets in developing countries as less attractive due to lower capacity to pay, higher transaction costs, weaker contract law and intellectual property rights, and general unfamiliarity.

Figure 7, drawn from UNEP's "Global Trends in Sustainable Energy Investment 2007" (UNEP 2007) shows the disparity of investments in sustainable energy between developed and developing countries. Although "sustainable energy" in this context refers only to renewable energy and energy efficiency (thus leaving out many CETs such as CCS), and the figure does not explicitly split out RD&D investments, the trend is very clear: The disparity between wealthy and poor countries in RD&D investments covering the full range of promising CETs is even larger.

Intellectual Property Right Protection is a Concern

The amount of R&D in any field is affected by the strength of intellectual property right (IPR) protection. Companies that invest heavily to develop new technologies must be guaranteed the benefits of their innovations. In most OECD countries, IPR protection is sufficient and—although not perfectly flawless on an international basis—this protection generally works well among countries within the industrialized world. However, the fear of substantially weaker IPR protection in many developing countries—and the corollary threat of technology theft—deters investment in this field. For some energy technologies, for which sophisticated technological advances and large sums of capital are required, this factor is particularly relevant. Strong IPR can improve incentives to develop such technologies. Yet, strong IPRs deter the adoption and diffusion of new technologies once they are commercialized. For other clean energy technologies the competition-limiting effects of IPR are less pronounced, especially if compared with other sectors, such as pharmaceuticals. In fact, the basic approaches

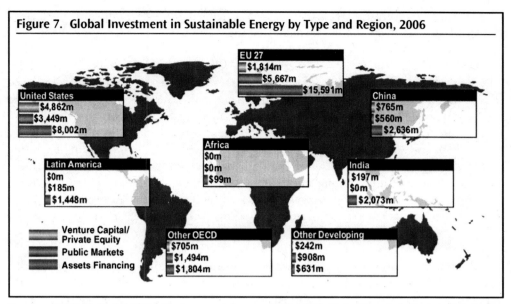

Figure 7. **Global Investment in Sustainable Energy by Type and Region, 2006**

Source: UNEP (United Nations Environment Programme). 2007. *Global Trends in Sustainable Energy Investment 2007: Analysis of the Trends and Issues of Renewable Energy and Energy Efficiency in OECD and Developing Countries.* ISBN: 978.92.807.289-0. Paris: UNEP Sustainable Energy Initiative (SEFI) and New Energy Finance Ltd. 2007.

to solving a technological problem have often long been off-patent for clean energy technologies and thus only specific improvements or features are protected by IPR. This allows for considerable competition between patented products (Barton 2007).

The Network Structure of the Electricity Sector Limits Integration of New Technology

Electricity sectors are organized around a model of large centralized power-generating stations that distribute electricity to end consumers along a vast network of transmission and distribution lines. With such a highly coordinated system, it is more difficult to introduce a new technology that does not fit well with the other existing components. For example, renewable energy systems with intermittent generation for example (such as wind and solar) face barriers being integrated into an electricity system that is designed around fully dispatchable generating units. Another example is distributed generation, whose potential can only be fully realized if allowed to sell electricity back into the grid, which most transmission and distribution networks are not equipped to receive.

National Interests Can Impede International Collaboration

International cooperation on clean energy technology development can effectively pool funding and share technical expertise, with each country contributing according to its area of expertise. However, governments, with their mandates to serve national interests, may be insufficiently motivated to participate in such cooperation. This impediment to clean energy RD&D can take different forms, including the following:

- *Unwillingness to share advanced technology.* Clean energy technologies may be a large market and technical advances that help commercialize clean energy products could become very valuable. Governments will naturally not want such advances that are developed by their own research or industries to be spread internationally without compensation.
- *Promotion of own industrial concerns.* Governments have an interest in promoting their own companies and industries. Thus, funding and other support will be less likely to go toward international efforts that support the industries of other countries.
- *Desire to control and take credit for funding.* By pooling funding from other countries, governments lose a degree of control over how money is spent. Governments will also have to share the credit for any advances that come from collaborative efforts.

While many international fora have been established to overcome national governments' reluctance to cooperate on clean energy technology development, to date these fora have very rarely extended to actual R&D efforts and focus primarily on information sharing and setting benchmarks.[5]

Energy RD&D Can Require Large, Sunk Capital Investments

The nature of the physical assets required to perform energy R&D can deter activity in the sector. The equipment needed for much of the energy R&D requires large, up-front capital investments. For example, the IEA estimates that a single CCS demonstration facility would cost between $500 million and $1 billion (IEA 2006a, 199). This investment makes much energy R&D possible for only a selected number of large companies. Also, the assets are highly specialized and have little, if any, value beyond research purposes, making them truly sunk costs. Thus, energy R&D is riskier than R&D in other sectors where the technical infrastructure would retain more of its value even if the research did not produce useful innovations. This higher risk leads to higher cost of capital and is thus a barrier for most energy RD&D.

The Commodity Nature of Electricity

Electricity is a homogeneous good that is indistinguishable in terms of its source or production technology. Consequently, electricity is traded on commodity markets and producers are not able to differentiate their product by quality and price. Electricity producers are therefore less able to extract price premia on innovative products from early adopters and quality-conscious consumers—a key source of R&D financing in other industries. Consequently, R&D in new electricity-generation technologies is undermined.

---

5. The ITER project (described in Appendix D) is an exception; the project has a combination of very high capital requirements and potentially high rewards, which, however, are still so uncertain and in the distant future that governments have been willing to share funding and risks.

"Carbon Lock-in," Subsidies, and Barriers to Trade

Current fossil fuel-based and carbon-intensive energy systems have benefited from long periods of increasing returns, creating positive feedbacks that reinforce the dominance of existing systems. This situation, termed "carbon lock-in," applies both to the technologies and the institutional structures that support them, creating significant barriers to the adoption of new technological alternatives (Foxon 2003).

Subsidies are one of the most obvious drivers. In fact, many countries have explicit or implicit subsidies for energy products. These subsidies are intended to favor certain consumer groups, support local energy production, or build political support. The IEA has estimated that in 2005, governments around the world provided $250 billion of subsidies to lower the price of energy to final consumers (IEA 2006b, 277–81). These subsidies cover energy products ranging from natural gas to oil products to electricity. Subsidies for established technologies deter RD&D because any new technology entering the market must compete against artificially low-priced alternatives. Also, the below-cost prices paid by consumers deter energy efficiency and thus the incentive to develop and deploy new energy-efficient technologies.

In addition, varied levels of tariff and nontariff barriers are impeding the diffusion of clean energy technologies in developing countries. For example, energy-efficient lighting in India is subject to a tariff of 30 percent and a nontariff equivalent of 106 percent (World Bank 2007b).

Imperfect and Asymmetric Information

Incentives to innovation are also constrained by the imperfect nature and uneven distribution of information between different innovators as well as between users and producers of technology. The development of new technologies is widely dispersed between different institutions and countries. Although competition is a key driver to innovation, limited information sharing due to concerns about loosing one's competitive edge and high transactions costs can also considerably slow down technological advancements. Limited opportunities for information sharing about future demands and future technological possibilities between technology users and producers also reduce incentives for innovation.

CHAPTER 5

Case Studies of Technical Innovation from Other Sectors

The barriers identified in Chapter 4 undermine the global development of clean energy technologies by reducing incentives to invest in costly RD&D and rendering RD&D efforts less effective. Thus, despite the recently increased engagement in this area by government and the private sector, significant gaps between public and private, as well as domestic and international RD&D efforts, will in all likelihood continue. The most important among the barriers discussed earlier—the uncertainty about future valuations of CO_2 emissions and the global nature of climate change—undermine incentives to perform clean energy RD&D for the private sector and national governments. Considering the rapidly growing global GHG emissions and the long lead times of new energy technology development, the world can not afford to wait until these problems have been addressed through international responses. Moreover, even when such responses are finally in place, other barriers such as the "valley of death" will continue to hamper clean energy technology development.

Consequently, innovative approaches are urgently needed to mitigate the barriers and accelerate the development of clean energy technologies. This chapter presents the following four cases studies that describe how barriers similar to the ones affecting clean energy technologies have been overcome in non-energy sectors using novel and innovative approaches:

1. The Consultative Group on International Agricultural Research (CGIAR) and the development of new and improved high yielding crop varieties in developing countries,
2. The Advanced Market Commitment (AMC) and the development of vaccines for diseases prevalent in developing countries,
3. The Human Genome Project (HGP) and advances in biotechnology, and
4. The concept of distributed innovation and innovations in the software and IT industries.

These case studies were chosen because they had considerable impacts on advancing technological development in non-energy sectors. As with most fields, energy RD&D professionals will focus on their own expertise when looking for models and new ideas. By linking to other sectors and examining their successful approaches to technical innovation, this paper intends to introduce new thinking to clean energy RD&D. The case studies were also chosen because of their parallels with clean energy. Although the industries discussed in the case studies and the various clean energy technologies have many differences, many of the types of barriers that have been overcome in these case studies are similar to the challenges to the development of new energy technologies. The case studies include examples of how international cooperation as well as public-private partnerships can successfully provide for unmet public goods, overcome the "valley of death," and develop technologies aimed at client country needs. And importantly, the first two case studies—CGIAR for agriculture and AMC for vaccines—demonstrate how the World Bank was able to leverage its strengths to foster technical innovation. Table 1 summarizes these case studies, followed by more in-depth descriptions.

Agriculture and the Consultative Group on International Agricultural Research (CGIAR)

The Challenge: Increasing Food Supply in the Developing World

In the 1950s and 1960s, international concerns grew about the adequacy of the world food supply and various studies forecasted famines in developing countries. Low crop yields and high vulnerability to environmental and climatic conditions were recognized as the major barriers to increased food supply. Major investments in agricultural R&D specifically for developing country food crops were identified as the key to improving this situation.

However, the public-good nature of agricultural R&D reduced incentives for these necessary investments. Seeds of improved crop varieties are nonexcludable and nonrival in consumption, which makes it difficult for the private sector to fully appropriate the returns from their research investments and seriously undermines private R&D efforts. In addition, international spillovers from R&D can also lead to underinvestment in the public sector (World Bank 2003; Pardy, Beintema, Dehmer, and Wood 2006). While intellectual property rights (IPR), large private markets, and effective government institutions mitigated these effects in developed countries, agricultural R&D in low-income countries was seriously insufficient.

The Response: International Cooperation on Agricultural R&D

Following extensive international consultations, it was agreed that the market failures preventing agricultural R&D investments in developing countries should be addressed by a concerted international response that pooled international expertise and funding. For this purpose, the Consultative Group on International Agricultural Research (CGIAR) was founded in 1971 with the objective of boosting agricultural productivity in developing countries through the development and deployment of new technologies. Based on the pioneering work of the Ford and Rockefeller Foundations, the CGIAR began to fund Inter-

Table 1. Summary of Case Studies of Technology Innovation in Non-energy Sectors

Program	Purpose/Rationale	Description	Results
Consultative Group on International Agricultural Research (CGIAR)	→ *Achieve sustainable food security in the developing world through scientific research* • Limited incentives for the private sector to engage in R&D aimed at developing country crops • Limited domestic R&D capacity in developing countries • International response needed	*International Agricultural Research Centers (IARCs)* • Research centers established in certain key developing countries • Each center focuses on improving yields of a local food crop, which is then further adapted to the conditions in different countries *Programmatic approach: Challenge Programs* • Time-bound research networks that target complex issues and require partnerships across institutions	• Success of IARCs in fostering Green Revolution widely acknowledged. • $460 million of annual funding mobilized from multiple donors • Subsequent broadening of IARCs mandate beyond increasing crop yields generated less clear success • Programmatic approach justified by increasingly complex nature of agricultural research and biotechnology, but too early to measure success
Advanced Market Commitment for Vaccines (AMC)	→ *Develop vaccines for diseases prevalent in developing countries* • Six million people a year die from diseases such as HIV/AIDS, malaria, and tuberculosis • Insufficient incentives for private sector to engage in costly R&D for vaccines • Many vaccines are caught in the "valley of death"	• AMCs are financial commitments by donors to subsidize future purchase of improved vaccines not now available, if appropriate vaccine is developed <u>and</u> it is requested by developing countries • Concept obliges donors to make their purchases (up to an agreed-upon price and volume) only if and when a vaccine is developed that achieves specified technical advances and if there is demand for the vaccine	• Establishment of scientific committee to assess promising vaccine for AMC pilot • Pilot AMC for pneumococcal vaccine has strong donor and industry support • Multidonor pledges totaling $1.5 billion made in February 2007 • Ongoing discussions with donors and industry to define terms, financing arrangements, and contractual obligations

(continued)

Table 1. Summary of Case Studies of Technology Innovation in Non-energy Sectors (*Continued*)

Program	Purpose/Rationale	Description	Results
Human Genome Project (HGP)	→ *Map the Human Genome* • Huge potential benefits for advancements in biotechnology, medicine, and other sectors • High complexity, costs, dispersed expertise, and spillover effects of research hamper commercial R&D	• Joint Program by U.S. Department of Energy and National Institutes of Health with international support • Loosely coordinated international consortium distributes research grants on competitive basis • Results published in publicly accessible database to facilitate follow-on innovations	• Program achieved target well ahead of schedule and generated multitude of follow-on innovations • $3.3 billion of funding mobilized • After basic research phase, competition with private sector catalyzed advances in applied technology • Public effort beyond basic research controversial
Distributed Innovation	→ *Generate innovation through inclusive collaborative approaches* • Limitations of conventional IPRs • Leverage expertise and commitment of diverse set of innovators	*Open Source Software:* • Source code (computer script) is free and freely accessible to be changed and improved by anyone • Product generated is a public good, that is, nonexcludable and nonrival in consumption *Commercial Innovation Networks:* • Cooperation of large and diverse set of companies on development of products and services is beneficial • Sophisticated contractual agreements align incentives of participants	• Highly successful open source products, such as Apache, Linux, Wikipedia • Innovation networks increasingly and successfully used in industry, such as in the development of the iPod • Limitations in terms of applicability to certain technologies and continuing concerns about IPR

national Agricultural Research Centers (IARC) in Colombia, Mexico, Nigeria, and the Philippines. These IARCs focused on research and development of high-yielding varieties of the major food staples in developing countries, each specializing on the agriculture of a particular region. Most importantly, the centers conducted agricultural research to generate new and improved germplasm that could be transferred and adapted to various developing countries. Their effectiveness was enhanced by the ability of the CGIAR to focus its funds and expertise on a clearly defined and relatively narrow set of technical goals: the improvement of yields for certain food crops

The CGIAR was founded as an umbrella organization for the IARCs and consisted only of a loose cooperative agreement without legal identity, charter, formal requirements for membership, or central decision-making body. Membership was and remains open to any country or organization that is willing to support the CGIAR commitment to a common R&D agenda. Decisions are taken by consensus while information and guidance on research focus is provided by an independent *Science Council.* The CGIAR relies on voluntary contributions from its membership base—a broad group of governments, private foundations, and international and regional organizations, including the World Bank.

In 2006, the CGIAR annual budget was $448 million (CGIAR 2007). This compares with total public agricultural research expenditures worldwide of $23 billion in 2000, of which $12.8 billion was spent in developing countries (see table 2). Private sector agricultural

Table 2. Agricultural R&D Spending and the Role of the CGIAR

Source	Agricultural R&D Spending (million 2000 international dollars)				Share (percent)	
	Public			Private	Public	Private
Year	1981	1991	2000	2000	2000	
Asia-Pacific	3,047	4,847	7,523	663	91.9	8.1
Latin America and the Caribbean	1,897	2,107	2,454	124	95.2	4.8
Sub-Saharan Africa	1,196	1,365	1,461	26	98.3	1.7
Middle East and North Africa	764	1,139	1,382	50	96.5	3.5
Developing country subtotal	6,904	9,459	12,819	862	93.7	6.3
High-income country subtotal	8,293	10,534	10,191	12,086	45.7	54.3
Total	15,197	19,992	23,010	12,984	64.0	36.0
CGIAR annual budget	242.6	357.7	331.0			
% of public developing country R&D	3.5%	3.8%	2.6%			
% of total public R&D	1.6%	1.8%	1.4%			

Sources: Pardey, Beintema, Dehmer, and Wood 2006; CGIAR *Annual Report,* various issues; Deflators from World Bank Development Indicators.

R&D only plays a small role in developing countries. Its share of total agricultural R&D spending is 6.3 percent on average and just 1.7 percent in Sub-Saharan Africa. This private sector share compares with 54.3 percent in the high-income countries. The CGIAR budget accounted for 3.5 percent of public agricultural research in developing countries in 1981, 3.8 percent in 1991 and 2.6 percent in 2000.

Since its origin in 1971, the CGIAR has followed two major approaches to agricultural R&D: the center-based approach, then the programmatic approach. The early years were dominated by the center-based approach in which IARCs followed a narrow focus on increasing crop yields of specific food staples with the ultimate goal of increasing the availability of food in developing countries. The technical challenges of these tasks were comparatively limited and research took place with relatively low capital intensity. From the original four IARCs in 1971, the CGIAR grew to a high of 18 centers in 1990, and down to 15 centers by 2007. The IARCs are independent institutions, each with its own charter, international board of trustees, director general, and staff. Technically, they focus mainly on conducting research and creating new germplasm from which to develop new, high-yield seeds. This germplasm is then made publicly available—CGIAR maintains and manages publicly accessible gene banks—and adapted to particular local climate and soils by National Agricultural Research Systems (NARS). These systems are composed mostly of local and national governments, research institutes, and universities that cooperate on the development of new seed varieties and their distribution to farmers.

During the 1980s and 1990s, the CGIAR mandate was expanded to include not only the economic well-being of the rural poor, but also the protection of biodiversity, land, and water. This expansion meant that the earlier specialized and technical focus on improving crop yields was to an extent replaced by a broader approach that included issues of natural resource management, policy, and local capacity building. At the same time, and independently of the CGIAR, agricultural research itself evolved into a more sophisticated, capital-intensive, and multilayered industry involving a wide variety of highly specialized institutions. The share of private sector R&D also increased considerably, although it remained predominantly focused on the needs of high-income countries.

In response to these developments, the center-based approach of the IARCs was complemented with a programmatic approach. This second approach was designed to respond directly to the major concerns of the global development agenda by creatively mobilizing resources to address major global or regional issues. This mobilization was achieved through the launch of *Challenge Programs,* a series of time-bound, independently governed research networks that target complex issues of major significance regarding the CGIAR goals and that require partnerships among a wide range of public and private institutions. Instead of conducting research in one of the IARCs, these virtual networks attempt to harness the expertise and resources of a multitude of institutions through the development of virtual research networks. Themes for Challenge Programs are generated through a bottom-up-approach that invites the global agricultural research community to submit potential ideas.

Evaluation of CGIAR R&D

Despite its small share in total agricultural R&D, the CGIAR achieved considerable success in boosting agricultural productivity in developing countries. In fact, the role of the

CGIAR is seen as pivotal in bringing about the Green Revolution (World Bank 2003; Gagnun-Lebrun 2004; Pardey, Beintema, Dehmer, and Wood 2006). This success can probably be attributed to the ability of the CGIAR to focus its funds and expertise on a clearly defined and relatively narrow set of technical goals: the improvement of yields for certain food crops. Relatively low technical sophistication and capital requirements allowed the IARCs to conduct highly relevant research and field trials, the results of which the NARS could readily apply. Close cooperation between IARCs and NARS facilitated knowledge transfers and ensured adaptation of the technologies to various local requirements and ultimately deployment of the improved seeds to farmers in developing countries. This success is documented by a number of evaluations that estimate returns on CGIAR germplasm research investments of 40–78 percent (World Bank 2003).

In contrast to this evident early success, the effectiveness of the CGIAR during later periods has been subject to more criticism. Various analyses point to the fact that the broadening of the CGIAR mandate to include a variety of softer and less technical goals diluted the organization's focus. According to an evaluation of the CGIAR by the World Bank's Operations Evaluations Department (OED), this broadening has led to a situation in which "its current mix of activities reflect neither its comparative advantage nor its core competence" (ibid.). The CGIAR Secretariat disputed this view. The move of donors away from unrestricted funding to be used at the discretion of the centers toward funding restricted to specific areas exacerbated this trend.

The CGIAR also faces new challenges externally. The dramatic changes in international agricultural research, characterized by the rapidly developing biotechnology sector, a rise in capital intensity, growing importance of IPRs, and increased private sector activities all question where and how the CGIAR can most productively use its limited resources. While the introduction of the Challenge Programs appears to be a promising response, their ultimate success will be determined by their ability to foster partnerships with other public and private institutions and to develop innovative approaches to manage IPRs.

World Bank Role in CGIAR

The idea of forming a consultative group to pursue agricultural R&D to serve the needs of developing counties originated in the World Bank. In the 1960s the World Bank often used consultative groups to coordinate international donor activities in selected countries and former World Bank President, Robert McNamara, pushed forward the idea of extending the consultative group model to the agricultural sector.

The World Bank was instrumental in developing the secretariat of the CGIAR and has housed its administrative infrastructure since its beginning in 1971. The World Bank's experience and reputation as an administrator of international collaboration and funding made it well-suited to this task. CGIAR Secretariat employees are World Bank staff. The technical expertise on agricultural science is drawn less from the World Bank; the UN Food and Agriculture Organization (FAO) of the plays the lead role in this regard.

In FY07, the World Bank contributed about $50 million to CGIAR activities, ($5 million for secretariat administration and $45 million for operations), which is 11 percent of the CGIAR's total annual budget and is funded through the grants budget of the Development Grants Facility (DGF). The World Bank's contribution is the largest of any CGIAR donor except the United States, which attaches restrictions to part of its funding.

Vaccines and Advanced Market Commitments (AMCs)

The Challenge: Diseases Affecting the World's Poor Countries Receiving Insufficient Attention

Advances in medical research and efforts to increase access to immunization and medicine have led to substantial rises in health standards and life expectancy in high-income countries over the past century. By contrast, few medicines and vaccines have been developed for the world's poorest countries and 6 million people die from diseases such as HIV/AIDS, tuberculosis, and malaria each year (Global Forum for Health Research 2004). Preventive treatment through vaccines is generally seen as the most promising response to many epidemic diseases, but for a variety of reasons development of such vaccines has lagged. Most importantly, insufficient R&D funds have been directed at diseases in developing countries. Only 3 percent of total health R&D funding is directed at diseases prevalent in low- and middle-income countries although they are home to 84 percent of the world's population. High-income countries attract 97 percent of health-related R&D funding while accounting for only 16 percent of the world's population (Global Forum for Health Research 2006).

This neglect has several causes, all connected to poverty and the limited ability of governments and individuals to pay for medicines and vaccines. In high-income countries, the development of new vaccines is traditionally financed through a mixture of public and private funds. The early phases of development, which consist mostly of basic research, attract little private investment due to technology spillovers and are therefore dominated by the public sector. However, in most developing countries these public activities are very small due to low government budgets and limited capacity and infrastructure.

In high-income countries, the bulk of private R&D funds are devoted to later stages of vaccine development when, based on advances in basic research, actual vaccines are developed. Typically, this process involves several steps from basic research, clinical testing, regulatory approval, production, and finally distribution—a process that can take up to 20 years and may entail many unsuccessful efforts before a vaccine is finally brought to market. The very regulated clinical development process makes medical research highly risky and capital-intensive. It is estimated that vaccine development costs from several hundred million dollars to $1.5 billion (Levine, Kremer, and Albright 2005). Consequently, private sector R&D is devoted only to diseases that promise markets large enough to recoup these high sunk costs.

Unfortunately, markets for vaccines—especially for vaccines against diseases prevalent in developing countries—are too small and risky (from a revenue perspective) to justify such large investments. Vaccines represent only a very small share of global pharmaceutical markets. While vaccine sales in developing countries account for about 50 percent of the total sales volume, their value is only 5 percent of the world market owing to the limited purchasing power of patients in developing countries. Vaccine sales account for only 1.5 percent of the global pharmaceutical market; developing-country vaccines make up less than 0.1 percent of the global pharmaceutical market (ibid.). Combined with the unpredictable uptake of new vaccines by governments, these products become too small and risky to justify commercial R&D on a large scale.

The Response: Advanced Market Commitments for Vaccines

Advanced market commitments represent one way to spur developments of new vaccines for markets that commercial interests do not consider profitable or attractive. AMCs are financial commitments by participating donors to subsidize the future purchase of a vaccine not yet available, if an appropriate vaccine is made available and if it is requested by developing countries. The concept obliges donors to make their purchases (up to an agreed-upon price) only if and when a vaccine is developed that meets certain eligibility criteria.

The donor commitments are made via a complex and legally binding contractual framework. This framework specifies the explicit target product profile (based on health impact in developing countries) that the new vaccine must meet to be eligible. For example, the contract could set a minimum level of efficacy for the vaccine (that is, the percentage of population that would be protected from the disease in question after taking the vaccine). The framework also sets the price at which the vaccines will be purchased and the total amount of funds available to support purchase. An AMC is not a purchase guarantee, because industry will only receive the subsidized price if its product meets the targeted standards and countries demand the product. The latter is gauged by not subsidizing 100 percent of the vaccine purchasing price and requiring countries to make a co-payment that will be linked to the post-AMC price of each vaccine. In addition, the AMC encourages innovation and competition between drug companies for example, by being sized and structured to last roughly 10 years to provide incentives for second and third entrants to the market and by making the post-AMC price a competitive factor for demand.

Once the AMC funding is depleted and the obligations made to the drug developer and manufacturer are fulfilled, a pre-agreed "tail pricing" arrangement will go into effect that obliges the drug developer to produce and sell the vaccine at or below the price that they established when they entered into the AMC agreement. This arrangement ensures that developing countries have an understanding of the long-term price and are better positioned to sustain coverage. The AMC establishes a continuing supply obligation for firms that may be fulfilled through, for example, technology transfers. However, firms are under no obligation to share trade secrets about their new product and will maintain full intellectual property rights.

In this way, AMCs reduce the financing risk and inflate the usually small size of vaccine markets in developing countries and thereby mobilize the ingenuity of a multitude of private sector entities for the development of vaccines for neglected diseases. At the same time, the price subsidy helps to ensure that the vaccine is actually affordable in developing countries once it becomes ready for widespread use. AMCs are intended to complement existing prevention, treatment, and research in these areas. Because the money is only disbursed for successful vaccine development, theoretically AMC does not divert funds from being spent on existing solutions to control disease.

Pilot AMC

The GAVI Alliance and the World Bank designed a public AMC for pneumococcal vaccines to demonstrate its impact on accelerating vaccine development, production scale-up, and introduction. Pneumonia is the leading infectious cause of childhood mortality worldwide. An estimated 1.9 million (or 19 percent) of the estimated 10 million child deaths each year result from pneumonia. Pneumococcal disease is the leading cause of these child pneumonia deaths, as well as the second-leading cause of childhood meningitis deaths. Pneumococcal disease kills more than 1.6 million people each year, including 700,000 to 1 million children under 5 years of age. An independent expert committee, with representation from developing and industrialized countries assessed six potentially vaccine-preventable diseases and recommended that pneumococcal disease be the target of the initial market commitment.

In February 2007 a consortium of donors pledged to support the pilot with $1.5 billion in commitments (see table 3).

Table 3. AMC Donors Commitments for a Pneumococcal Vaccine	
Donor	**AMC Contribution**
Italy	$635 million
United Kingdom	$485 million
Canada	$200 million
Russian Federation	$80 million
Norway	$50 million
Bill & Melinda Gates Foundation	$50 million
Total	**$1.5 billion**

Source: www.vaccineamc.org

The AMC for pneumococcal vaccines is intended to offer an improved market for vaccines now in development. Vaccines are bought with donor-committed funds only if they meet predetermined standards of efficacy and safety. The price paid for the vaccine will be about $5–7 per dose. The government of each country to receive the vaccines will be expected to pay some share of this price. The exact amount of the government contribution has yet to be determined, but it is expected to stay small, at about $0.50 per dose. The AMC funding should be depleted in 7–10 years. Pilot participants expect pneumococcal vaccines to reach developing countries by 2010, or about 10 years earlier than without the AMC program. It is expected to save 5.4 million lives once the vaccine is developed.

AMC Evaluation

AMCs are a creative approach to fostering innovation. They harness the efforts of a multitude of private sector actors and liberate the donors from having to pick technological and commercial winners. Moreover, donor funding is only used after a qualifying vaccine has actually been developed and demanded. However, while it is too early to judge the success of AMCs, several observations can be made about certain key issues that may influence the ultimate success of the concept.

The biggest challenge to successful AMCs is properly devising the legally binding contracts on which the commitments are based. The commitments have to be credible to industry to be effective. Vaccine developers must be entirely certain that the money will be available. However, it is difficult for governments or other donors to make credible commitments 10 or more years in the future and firms are not enthusiastic about managing a financial relationship with seven or more sovereign governments. In addition, credibility is further undermined by the inconsistency of donors' incentives that will change over time. While donors have an incentive to make large funding promises *before* a vaccine is developed, governments and the public have an incentive to push for low prices and a wide distribution *after* a vaccine has been developed, thus potentially reducing the profit potential for the producer. To reduce transaction costs and increase the credibility of funding, donor governments are making hard commitments to the AMC, with some providing cash upfront, and have requested the World Bank to explore acting as the financial intermediary, guaranteeing funds.

Complications also arise from the need to design accurate and plausible contractual terms years before the vaccine is available or, in some cases, even developed. The greater the time frame and scientific and technical uncertainties (for example with early-stage products), the more difficult it is to set market parameters such as payable price.

Because it is a market mechanism, many have questioned how far back in the pipeline an AMC can reach: Is the "pull" great enough to impact investments in candidate technologies before proof of product is established? Adjusting the AMC market size for risk and time reduces the pull impact significantly on a product that is 10–20 years and many technical hurdles away. By one estimate, every $1.00 of market pull incentive leverages only $0.06 of additional R&D expenditures (Finkelstein 2004). Generally, market pull incentives are believed to solicit R&D devoted primarily to products at later stages of development. A second AMC pilot for an earlier stage vaccine is under consideration and may help test this question. In any case, the AMC is a complement rather than a substitute for classical R&D push policies.

Biotechnology and the Human Genome Project (HGP)

Deciphering the Human Genome

A genome underlies every aspect of human, animal, and plant life and health. It describes the entire DNA in an organism, including its genes and determines how an organism works and reacts to environmental factors. Understanding the effects of DNA variations among individuals, the functions of specific genes, and their interactions among themselves and with environmental factors is critical for diagnosing, treating, and potentially preventing diseases. In addition, information about genetic codes of other organisms such as animals and plants can contribute to solving challenges in agriculture, energy production, and carbon sequestration.

Deciphering genomes is highly complex. The human genome in particular consists of 3 billion base pairs, the decoding of which poses immense challenges for biology, chemistry, and medicine. At the same time, challenges in informatics and engineering have to be overcome to develop the machinery and analytical tools that allow visualizing, analyzing, and categorizing the genome. Consequently, genetic research requires the collaboration of a wide variety of experts from different disciplines. Each field must generate innovations that work in conjunction with each other to lay the foundation for deciphering the human genome.

The Human Genome Project

With the aim of meeting this challenge, the U.S. Department of Energy (DOE) and the National Institutes of Health (NIH) created the Human Genome Project (HGP) in 1990. Its goal was to generate a high-quality reference DNA sequence for the human genome and identify all human genes. The results would then be published in a publicly accessible database and thereby facilitate private and public innovation in a variety of fields. Because a project of this complexity and scale had never been undertaken in this area, the HGP had to develop an organizational structure that would unite experts from a variety of disciplines cooperating toward a common goal.

James Watson, the Nobel Prize-winning biologist who was instrumental in setting up HGP and served as its director from 1990 to 1993, put particular emphasis on broad involvement of the international research community. He deemed this international approach essential to tap into the best expertise as well as to allow the results of the HGP to be used as widely as possible and promote follow-on innovations. Thus, in addition to the two principal project sponsors from the U.S. government, other international partners were included in the HGP. The Wellcome Trust, a large private charitable institution focusing on human health in the United Kingdom, joined the effort as a major partner in 1992 (see table 4) and scientists in Canada, China, France, Germany, Israel, and Japan contributed to the research.

From an organizational perspective, HGP for most of its project life was that of a loosely coordinated international consortium. Although HGP set up an intramural research program, the majority of research was undertaken in external research centers. Six initial centers were established in the United States, all at universities: The Whitehead Institute for Biomedical Research, affiliated with MIT; the University of Michigan; Baylor College of Medicine; University of Utah; University of California, San Francisco; and Washington University in St. Louis. In addition, several other smaller centers in different countries contributed to the research.

Grants were distributed to all participating research centers on a competitive basis following a peer-review process in which submitted proposals were evaluated on their quality and relevance for the overall goal of the HGP. Thus, universities and other research institutions were granted relative creative freedom in their approaches as long as they contributed to the HGP goal. NIH, which provided the bulk of research funds, provided overall leadership, albeit on an informal basis. Uniting the various researchers at the different centers was the mutually agreed requirement that all participants had to place their findings into a common repository maintained by NIH, the GenBank. New information was required to be published at the GenBank within 24 hours

Table 4.	Human Genome Project Funding (US$ millions)			
Year	DOE	NIH	Wellcome Trust	Total
1988	10.7	17.2		27.9
1989	18.5	28.2		46.7
1990	27.2	59.5		86.7
1991	47.4	87.4		134.8
1992	59.4	104.8		164.2
1993	63.0	106.1		169.1
1994	63.3	127.0		190.3
1995	68.7	153.8		222.5
1996	73.9	169.3		243.2
1997	77.9	188.9		266.8
1998	85.5	218.3		303.8
1999	89.9	225.7		315.6
2000	88.9	271.7		360.6
2001	86.4	308.4		394.8
2002	87.8	346.7		434.3
1992–2000			306.0	
Total	948.5	2,066.3	306.0	3,320.8

Source: Lambright 2002.
Note: Only a fraction of the total funds was directly spent on deciphering the human genome. The majority of funds was spent on basic research developing technical and analytical tools and on deciphering the genomes of simpler organisms. In addition, the DOE and NIH genome programs set aside 3–5 percent of their respective total annual budgets for the study of the HGP's ethical, legal, and political issues.

of discovery and to be open to the public. NIH's role focused on developing the technical capacity to gather the resulting huge datasets and aggregating them in a meaningful pattern.

This work involved undertaking groundbreaking research in the areas of computer software and laboratory equipment. Much of this expertise was developed while researching the genomes of less-sophisticated organisms. In addition to the technical research, HGP also involved social scientists, ethicists, and legal scholars who were responsible for analyzing the social, ethical, and legal impacts and implications of genetic research.

Cooperation and Competition: Relations with the Private Sector

HGP was designed with the specific goal of putting information on the human genome in the public domain and thereby facilitating follow-on innovations by universities, government laboratories, and the private sector worldwide. From the early stages, HGP established relationships with private companies—as developers and providers of laboratory equipment, and as users of the information generated. In fact, technologies were licensed to private companies early on and grants were awarded for innovative research. In this way, the project catalyzed the multibillion-dollar U.S. biotechnology industry and fostered the development of new applications, especially in medicine.

However, while HGP entertained constructive relationships with a number of companies, the most defining relationship was the rivalry with Celera, a biotech company backed by venture capital. Celera was founded by J. Craig Venter, a former NIH researcher, with the aim of deciphering the human genome by using an innovative—albeit controversial—method know as the shotgun approach. Most NIH researchers did not take this approach seriously until 1995 when Venter succeeded in decoding the genome of *H. Influenzae.* In 1998, Venter unleashed a competitive race with HGP after he announced that his new company would be able to beat HGP in deciphering the human genome more quickly and less expensively by relying on his approach and on a new generation of sequencing machines. This announcement gave rise to a race to decipher the human genome that was closely watched and discussed in public.

Francis Collins, the new HGP director, understood that the well-financed biotech venture had an advantage over HGP—not least due to the venture's compact organizational structure that allowed it to adapt quickly to new challenges. As a reaction, from 1998 to 2001 Collins centralized HGP's structure, shifting the great majority of the research to a core set of research centers over which he exerted more direct control: the U.S. DOE Joint Genome Institute, the Baylor College of Medicine Genome Sequencing Center, the Wellcome Trust Sanger Institute, the Washington University School of Medicine Genome Sequencing Center, and the Whitehead Institute/MIT Center for Genome Research.

In June 2000, closely watched by the public media, scientists announced the completion of the first working draft of the entire human genome; in a White House ceremony both HGP and Celera were declared as "winners" of the genome race. First analyses of the details appeared in the February 2001 issues of the journals *Nature* and *Science,* in which HGP and Celera separately described their findings. The high-quality reference sequence was finally completed in April 2003, marking the end of the Human Genome Project—two years ahead of the original schedule.

Assessment

The benefits of having decoded the human genome are undisputed: Publicly available to researchers worldwide, the human genome reference sequence provides a resource that serves as a basis for research and discovery in wide variety of applications. However, in addition to ethical, legal, and social issues associated with genetic research that are now being widely discussed, HGP raises important questions regarding the role of the public sector in scientific research. While the answers to these questions are too complicated to be comprehensively discussed here, the following broad points can be made.

First, it is apparent that government engagement was necessary to raise public awareness of the scientific potential of decoding the human genome, to perform the related basic research, and to develop the first analytical tools. In fact, Celera itself benefited from these innovations not only directly as inputs to their research but also indirectly, by giving publicity to the issue and thereby facilitating access to private finance. In fact, a number of earlier efforts by private firms to secure financing on similar ventures had failed, in part because prior to the HGP's initial successes, the work was seen as excessively risky.

Second, the ultimate success in decoding the human genome required an evolution of the project's organizational structure as early successes led to more advanced work. The complexity of these early stages of development required the involvement of experts from a variety of fields. Centers of excellence for these issues are widely dispersed nationally and internationally. Consequently, HGP in its early years adopted a loose, cooperative, and decentralized organizational structure that served the basic science stages well and provided the basis for broad and international follow-on innovations. However, this loose structure was far less efficient further down the innovation chain and it is improbable that HGP would have adapted its structure without the competition from Celera and the resulting adoption of a more compact organizational structure. This fact raises the question of whether HGP should even have continued to stay engaged in genome research after it had laid the groundwork on which the private sector could build.

As demonstrated by Celera's success, the private sector was able to take the research forward quite efficiently. But open access to HGP's innovations was a precondition for the success of Celera and numerous other companies. Thus, apart from ethical questions regarding intellectual property rights in relation to the human genome, efficiency considerations in conjunction with follow-on innovations may well have warranted public financing of the HGP. The strategy of collaborating with a variety of private sector actors, granting them open access to the deciphered genome, as well as licensing related technologies served the cause of technical innovation well in this case.

Open Source Software, Creation Networks, and Distributed Innovation

The Concept of Distributed Innovation

Innovation networks (or distributed innovation) has been receiving much attention in the last 10–15 years as a result of open-source products such as Linux and internet-based collaborative sites such as Wikipedia. Innovation networks are processes of technological development where large numbers of dispersed persons contribute toward innovation cooperatively, using nontraditional forms of intellectual property rights (IPR). Often contributors are not restricted to researchers in the traditional sense but include everyone, including consumers.

The concept is closely linked to the rise of the internet as an inexpensive and efficient means of communication—without it, collaboration between large numbers of widely dispersed people would be prohibitively expensive and inefficient. Distributed innovation has two main approaches: (i) a completely decentralized, open, and nonprofit process such as the one exemplified by open source software and (ii) a commercial creation internet process that limits participation and IPRs to invited members of an otherwise closed consortium.

Open Source Software and Internet Wikis

The more prominent approach to innovation networks follows a bottom-up, fully decentralized, and noncommercial methodology. The product generated by this process is freely accessible, completely nonrival in consumption, and nonexcludable (that is, a public good in the traditional sense). In accordance with economic theory, public goods tend to be underprovided but the open source process has proven successful on many occasions in overcoming this problem.

The essence of open source software is that source code (the computer script that determines the operation of a program) is free and freely accessible. Open source software is distributed together with its source code, usually under an open source license that:

- allows the free redistribution of the software without royalties or licensing fees to the author,
- requires that source code be distributed with the software or otherwise made available, and
- allows anyone to modify the software or develop other software from it, and to redistribute the modified software under the same terms.

Thus, turning the traditional notion of IPR upside down, open source software specifically encourages and facilitates the modification, adaptation, and free distribution of software products by anyone with the interest and expertise to do so. The operating system Linux and the online encyclopedia Wikipedia are among the most successful applications of this concept.

Linux. Linux is probably the most successful example of free and open source software. Its first components were developed by Linus Torwalds in 1991 and later integrated with software components from the nonprofit GNU project to create an integrated computer operating system. Linux was openly put on the internet together with its programming code to encourage testing, adaptation, and further development of the operating system and applications by a wide range of people.

While the programming and debugging of the software initially proceeded completely decentralized, the huge success of Linux and the growing numbers of contributors called for streamlining the process to allow the organization of the many comments and changes to the source code. Thus, a rather hierarchical though informal system evolved by which

Linus Torwalds and a group of so-called lieutenants—programmers with responsibility for specific subcomponents—organized the versions and picked and integrated the most useful contributions to the code.

This system has been highly successful. It has a huge advantage in the area of debugging, where the large number of test runs through the open source process allow it to outperform its main competitor, Windows NT, which relies on in-house debugging. Linux is widely used as an operating system for a variety of IT hardware, including servers, supercomputers, and mobile phones. Linux has gained the support of corporations such as Dell, IBM, Novell, and Sun Microsystems. According to market research company IDC, 25 percent of servers and 2.8 percent of desktop computers ran Linux as of 2004.

Wikipedia. Wikipedia is a multilingual free online encyclopedia, operated by the nonprofit Wikipedia Foundation that was directly inspired by the open source software movement. Wikipedia allows every individual to create and edit online articles; through this process of multiple edits by consumers and peer reviewers, an extremely fast growing encyclopedia has developed.

Due to Wikipedia's open nature and limited access control, critics have questioned its reliability and accuracy as well as the possibility that unknown or unidentifiable contributors would purposely post false information. Although multiple successive edits most often correct such misinformation, the process also may tend to dilute fact-based information toward the lowest common denominator on which a consensus can form.

Nevertheless, Wikipedia's success is undisputed and it currently ranks among the top 10 most-visited websites worldwide. As of December 2007, Wikipedia had approximately 9.3 million articles in 253 languages, 2.1 million of which are in the English edition. This volume makes it the largest, most extensive, and fastest growing encyclopedia ever compiled.

Commercial Innovation Networks

Commercial innovation (or creation) networks[6] (or nets) aim at leveraging the benefits of allowing a large and diverse set of people to share information and cooperate on the development of profitable products and services. Specifically, it is intended to overcome the "not invented here" syndrome that often hampers innovation within large research institutions and has often led large firms to miss new trends, such as the internet.

Commercial creation nets have different approaches; the main differences relate to levels of openness and hierarchical structures. Some creation nets are limited to a specific firm or institution and are supposed to facilitate cooperation between different departments within a single company or institution. For example, anyone in a firm could easily comment on specific product development by the research department and propose changes or wholly new ideas. The Dutch Bank ING has developed such an in-house network.

Other creation nets are cast more widely and specifically aim at fostering greater information sharing and cooperation between a consortium of companies, for example between a technology producer and its suppliers. Some approaches follow a completely open source approach *within the net,* whereby each participating company has access to and is able to

---

6. Similar concepts are also known as targeted innovation and targeted procurement.

propose changes to, for example, technology interfaces. At the same time, each participant retains a clearly specified role and responsibility for the delivery of specific components.

Apple's iPod is one famous example of a successful product developed through a creation network. Apple outsourced the design of the iPod system that integrates hardware, software, and music distributors to a consulting firm that pursued its development in the form of a creation network. Apple's own researchers concentrated on aggregating and packaging the final product. The benefits of this approach were a combination of speed and creativity almost unthinkable within the boundaries of a single firm; largely as a result of this approach, the iPod was developed in an astounding eight months. Similar approaches are successfully being pursued by Taiwan-based original-design manufacturers such as Lite-On Technology and Compal Electronics, and increasingly by large multinationals such as Procter & Gamble.

While creation nets are open with regards to the freedom to innovate, they require strong and often rigid institutional mechanisms that define modes of participation, dispute resolution procedures, and performance measurement. These structures also have to define how IPRs are managed within the network.

Assessment

Decentralized innovation—in both of the forms described above—provides many examples of successful product development. In both open access and for-profit applications, the power of involving large numbers of people to solve problems and aggregate information has generated impressive results. Importantly, the innovative potential is not only determined by the number of people who can be leveraged, but also by the degree of diversity of participants, which is almost never seen in traditional research processes.

Open source nonprofit approaches have been able to prove that under certain circumstances, they can generate high quality public goods, even when standard economics would suggest otherwise. In particular, such approaches are most successful for products that can be disaggregated into small, parallel tasks. Software programming is one such example. However, the process has proven less productive at generating completely new ideas for products than in optimizing or further developing existing ones. Inventions require key innovators who originate and develop an initial idea that is only later advanced forward in a cooperative way. Open source approaches are also limited by how they mobilize researchers to contribute. Research on open source software indicates that key motives for participants are peer recognition, solving practical issues that provide immediate benefits to the researchers, and contributing to a greater good or ideal. In addition, economists argue that benefiting from "learning by doing" as well as signaling effects to potential employers can be important motivating factors.

The fields in which network innovations have been carried out remain relatively small and the majority of technologies have not (yet) been able to induce innovation by such methods. Most technical and other innovation is still produced as a private for-profit good, protected by IPRs. In particular, technologies that require high capital investments and specific technical tools and methodologies are believed to fall into this category. Commercial creation networks that allow leveraging talent from a wider base offer a promising approach for accelerating innovation in such industries.

CHAPTER 6

Lessons Learned

The four case studies described in Chapter 5 offer important examples of how novel approaches can overcome barriers to technical innovation. Although each of the case studies examined has its own set of circumstances that differ from the clean energy sector, the similarities and the creative approaches used in RD&D can provide valuable lessons. This chapter lists seven sets of lessons.

Bridging the "Valley of Death"

The case studies offer different models for overcoming the "valley of death" using innovative institutional set-ups and financing mechanisms. AMC's promise of funding for *future* products not yet available effectively motivates the pharmaceutical industry to go further up the innovation chain than they would normally consider profitable. The structure shields donors—most of whom are also engaged in directly supporting basic pharmaceutical R&D—from the technical and commercial risks associated with vaccine development because donors will pay nothing if the desired vaccine is not developed. Instead, incentives are created for the private sector to cover these risks for which it is much better suited.

In the case of HGP, public R&D "supply-push" funding was provided to advance innovation past the valley of death. At the same time, private sector participation was encouraged from the early stages through the establishment of relationships with private companies both as providers of laboratory equipment and as users of the information generated. Technologies were also licensed to private companies and some grant money was made available, thus catalyzing the biotechnology industry and leading to new applications of the basic research, especially in medicine.

The CGIAR IARCs had a heavy "push" component with government funding for basic research. However, they also included product demonstration, local outreach, and capacity building to ensure the new products were best suited by the private actors who used them (that is, seeds for farmers) and that uptake of these new products could occur as seamlessly as possible. Distributed innovation offers another paradigm for overcoming the "valley of death." Segmenting the steps required for technical progress into more manageable pieces allows more entities both from the public and the private sectors to contribute to technology development in the gap that normally falls between the sectors. *The cumulative effect of these smaller steps of progress coupled with active public-private partnership can be sufficient to get over the "valley of death."*

Pooling Resources to Address Global Public Goods

Both CGIAR and AMC rely on pooling international donor funds. This collective action leverages individual governments' and other donors' contributions to enable increased effectiveness in addressing issues of public good. Distributed innovation also uses pooled contributions from around the world, but instead of gathering funds, it gathers expertise. A single researcher working on a small part of a larger puzzle is substantially more likely to solve the puzzle with the knowledge that many others are working with him or her to produce a collective product that all can later use. AMC solves the problem of sustaining private sector commitment over the long period of RD&D challenges by promising future money for technologies serving the public good. This monetary enticement—only delivered if industry properly responds—drives private companies to devote resources and expertise to develop technologies that serve the public good. *Establishing pooled, coordinated resources to support clean energy technologies would also improve the effectiveness of donors' limited resources.*

Facilitating Innovative Research Partnerships

The case studies demonstrate that RD&D structures that promote nontraditional partnerships focused on a clearly defined mission can effectively accelerate technology development. Partnerships that facilitate contributions from diverse research groups that would not normally collaborate allow the development process to tap into more areas of expertise, including public-private partnerships and international collaboration. CGIAR's Challenge Programs require partnerships among a wide range of public and private institutions and are intentionally designed to create virtual networks that harness the expertise of a multitude of different institutions. The Human Genome Project put in place a sophisticated contractual framework explicitly linking private companies to traditional research work. In addition, HGP put particular focus on the broad involvement of the international research community. Although the two principal sponsors were from the U.S. government, the structure of the loosely coordinated international consortium allowed essential contributions from numerous other countries. In addition, the open nature of its central data repository facilitated follow-on innovations worldwide. The structure of distributed innovation, such as open source software, intentionally breaks down the barriers to participation from any entity—public or private from any country in the world—to contribute according to its area of expertise. *Research programs that create the platform to encourage nontraditional partnerships can tap into a greater wealth of technical expertise.*

Transferring Technology: South-South and North-South

These case studies provide a set of models on how to marshal resources and expertise from the OECD to serve the technology needs of developing countries. For example, the center-based approach of CGIAR showed how establishing physical research centers—supplied largely by OECD funding and expertise—in developing countries resulted in new technologies (for example, seed varieties) suited both to local need and conditions. Moreover, this model led to successful South-South cooperation because centers from different countries shared experiences among themselves. In addition, the CGIAR center model facilitated capacity building in the host countries through its physical presence, enabling the better follow-through for technology introduction.

The AMC, on the other hand, leverages funding and expertise available in the North to develop vaccines needed in the South. This "pull" mechanism, by creating an incentive for the private sector at the end of its research, is essentially different from the "push" mechanism of CGIAR, where the research is directly funded by the donor organization. The structure of each industry and the nature of the technology to develop determines which approach is suitable for technology transfer. *Technology transfer can be best facilitated through active involvement both of developed and developing countries focusing on a joint ambition.*

Sharing Information and Addressing Intellectual Property Rights

These case studies demonstrate how active and open approaches to information sharing that still address the IPR concerns of the contributors help to assist technology development. The best example of this approach is the case for distributed innovation in software development, which is built on the paradigm of facilitating information exchange between diverse stakeholders. Because all results from the collective work are mandated to be free and freely accessible to all, contributors are not worried that they will lose access to any part of the combined work product, and thus contribute freely. Related concepts are found in the HGP and CGIAR. The central data repository of the Human Genome Project was deliberately designed to coordinate the work between different research centers engaged in deciphering the human genome and—at the same time—facilitating follow-on innovations by the private sector. The CGIAR strongly promotes information sharing between its research centers to adapt innovations to local conditions and allows broad access to these innovations. Innovative approaches to information sharing and IPR are also increasingly—and with growing success—used by the private sector. *Open access to nonproprietary information, agreed at the initial stages of development, will be particularly helpful in accelerating the research phase of new technology development.*

Setting Goals without Picking Winners

The case studies provide examples of how policy makers can set a clearly defined goal with-out constraining the technical avenues for achievement. In this way, they demonstrate how the proper degree and type of influence over research is essential to meet larger policy goals while fostering innovation. For the AMC, the decision of which vaccine to pursue was based on the recommendation of a politically independent expert committee consisting of public health specialists, economists, vaccinologists, and others. A second expert panel then set the required "target product profile" for AMC funding eligibility. After these deci-sions were made, the technical means of producing the vaccine were left to the industry. CGIAR's Challenge Programs puts even more decision making on the researchers. The themes from the Challenge Programs are generated through a bottom-up approach that invites agricultural researchers from around the world to submit ideas for potential goals most worthy of pursuit. The goal-setting for distributed innovation was wholly different because market forces set the goal for the researchers. In the case of distributed innovation in the software and high-tech sectors, defining research priorities is even more decentralized and is often entirely left to the market. *Policy makers should be as collaborative as possible in setting the general goals of the research activity while leaving the technical means of achieving it entirely to researchers.*

Using World Bank Group Strengths to Promote Technology Development

Both CGIAR and AMC show how strengths of the World Bank Group can be instrumental in developing innovative RD&D vehicles to promote important technology development not being pursued by traditional public and private actors. This is particularly the case for the technology needs of client countries. For CGIAR and AMC, the World Bank was able to use its convening power to bring together different stakeholders with a high level of technical expertise and to attract funding from multiple sources. Its strong analytical capability—and credibility in this area—provided an important segment of the intellectual thinking that supported these two initiatives. *The World Bank's experience in coordinating donor contribu-tions in novel and complex ways was essential in facilitating the operation of the novel RD&D structures in both of these cases.*

Table 5 provides an initial assessment of how the innovative RD&D approaches in the case studies apply to the major barriers to clean energy technology development. Adapting and applying these lessons to develop innovative vehicles to accelerate the development of clean energy technologies holds great promise and should therefore be pursued.

Table 5. Innovative RD&D Approaches to Address Barriers to Technology Development

Barriers to Clean Energy RD&D	Models of Innovative RD&D from Other Sectors				
	CGIAR		Advanced Market Commitment (AMC)	Human Genome Project (HGP)	Distributed Innovation
	IARC	Programmatic			
Negative externality of carbon emissions is difficult to valuate	Traditional "push"[a] to research technologies for which there was no adequate market "pull"[b]	Traditional "push" to research technologies for which there was no adequate market "pull"	Creates "artificial price" to replicate the externality not valued by market, creating price and quantity floor if product sells	n.a.	n.a.
Climate change mitigation is a global public good	Coordinated agreement of multiple donors to commit resources and expertise to pursue unmet public good	Coordinated agreement of multiple donors to commit resources and expertise to pursue unmet public good	Coordinated agreement of multiple donors motivates industry to serve unmet public good	Individual government initiative provides global public good (some international contributions)	Open source software and wikis share results of collaborative work equally and freely
The "valley of death" and the "mountain of death" impede technology development	Basic research linked to local outreach, product demonstration, and capacity building	Challenge Programs provide opportunities for joint ventures with the private sector	Creates "pull" mechanisms on the other side of "valley of death" to entice private sector involvement	Public research specifically closes R&D gaps and induces early and expansive private sector involvement	Segmentation of innovation progress increments and novel information sharing allow contributions from multiple non-traditional entities

(continued)

Table 5. Innovative RD&D Approaches to Address Barriers to Technology Development (Continued)

Barriers to Clean Energy RD&D	Models of Innovative RD&D from Other Sectors				
	CGIAR		Advanced Market Commitment (AMC)	Human Genome Project (HGP)	Distributed Innovation
	IARC	Programmatic			
Technology needs of developing countries are not adequately served	Research conducted relevant to developing country needs in partnership with developing country organizations	Themes for Challenge Programs directed toward developing country needs	Motivates OECD industry to pursue technologies to serve identified developing country needs	n.a.	n.a.
Intellectual Property Right (IPR) Protection	All research results are available to all and are actively disseminated internationally	Innovative IPR and knowledge sharing structures create incentives to participate	IPR of product developer is protected (unless production obligations are unmet)	All research results are shared openly	Innovative IPR and knowledge sharing structures create incentives to participate
National interests can impede international collaboration	Up-front agreement on international cooperation forms basis for vehicle	Up-front agreement on international cooperation forms basis for vehicle	Up-front agreement on international cooperation forms basis for vehicle	One country with significant outreach to others leads project	Open structure invites contributions from diverse actors who share goals and incentives, irrespective of nationalities

Note: n.a. = Not applicable.
a. "Supply push" describes efforts by governments to directly fund or facilitate R&D in specific research centers or topic areas.
b. "Demand pull" describes efforts by governments that increase the potential market for new products and thereby attract private sector R&D investments to cater to the market.

CHAPTER 7

Going Forward

This paper demonstrates that balancing climate change mitigation and increased energy needs in developing countries poses a serious dilemma that can only be reconciled with new and improved clean energy technologies. Unfortunately, despite renewed activity by governments and industry, serious barriers to energy RD&D will continue to undermine progress in this area. To introduce new thinking in addressing these concerns, this paper has examined four cases from outside the energy sector where creative approaches to RD&D have successfully overcome similar barriers. By looking outside the realm of traditional and ongoing energy RD&D this paper introduces several novel paradigms that have had substantial success in delivering new and useful products. The lessons learned from these case studies provide global cross-sectoral knowledge and experience transfer that introduces a range of examples about how RD&D could be approached for clean energy.

Specifically, two vehicles could be envisioned that build upon the previously discussed lessons learned:

A clean energy technology innovation network would virtually connect energy research centers in developing countries and support designated centers of excellence in providing a variety of services and financing tools to promote innovation. The centers could focus on particular technologies and serve as conduits to launch international challenge programs that finance R&D on specific issues.

An advanced purchase commitment for clean energy technologies. An independent scientific committee would specify the characteristics of a technology or portfolio of technologies, the commercialization of which would be rewarded by a future purchase commitment. Similar to an options contract on a future feed-in tariff, such a vehicle could be designed to provide incentives to commercialize technologies that currently languish in the "valley of death."

More research is required to define the structures that would make these vehicles most effective and efficient, as well as to determine which technologies should be targeted.

APPENDIXES

The Stages of Energy Technology Innovation

Technology innovation is a complex process influenced predominantly by the potential market for the technology, investment in R&D, spillovers from other technology areas, and the general advancement of science. For illustrative purposes (see figure A1.1), this process can be divided into four stages—research and development (R&D), demonstration, scale-up/deployment, and commercialization—each of which is characterized by different technical and institutional barriers. Consequently, the private and public sectors play different roles along the innovation chain.

The early phases of technological innovation focus on basic research to find solutions to specific technical problems. During the development phase, the research findings are applied to new technologies and products. Demonstration projects are then undertaken to further adapt the technology and demonstrate the functioning in larger-scale and real-world applications. Because of elevated technological risks and technology spillovers that prevent private companies from fully capturing the commercial benefits, the earlier stages of technology development are usually funded predominantly by public sources. After fundamental technical barriers have been resolved and the commercial potential of a technology becomes apparent, private sector funding becomes prevalent. Private funding increases during the scale-up phase, in which—most often with public assistance—technology is deployed on a larger scale. Private participation increases through full commercialization as public involvement drops off.

It should be noted that the junctions between these stages are ill-defined and characterized by multiple dynamic feedbacks. The advancement of innovations from one stage of development to the next is not automatic: the majority of innovations in any stage fail. Moreover, the transition from predominantly public research to predominantly private product development is not smooth and funding gaps between the various stages often prevent technically proven ideas from reaching large-scale deployment. In some technology

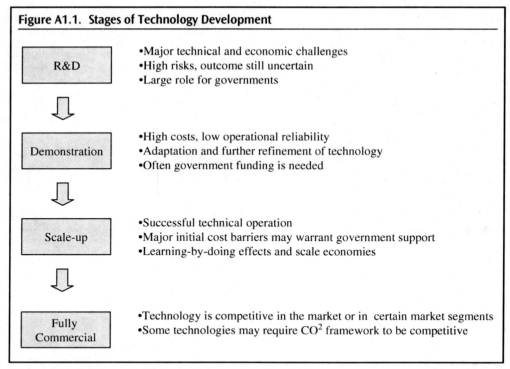

Figure A1.1. Stages of Technology Development

R&D
- Major technical and economic challenges
- High risks, outcome still uncertain
- Large role for governments

Demonstration
- High costs, low operational reliability
- Adaptation and further refinement of technology
- Often government funding is needed

Scale-up
- Successful technical operation
- Major initial cost barriers may warrant government support
- Learning-by-doing effects and scale economies

Fully Commercial
- Technology is competitive in the market or in certain market segments
- Some technologies may require CO^2 framework to be competitive

Source: Adapted from IEA 2006a.

sectors, active involvement by the venture capital industry plays an important role in closing some of these gaps for the most promising innovations. Similarly, many governments support the development of commercial products generated from basic R&D through a variety of measures. For example, government-supported business incubators that provide financial, management, and technical support to young technology ventures play an important role in pushing innovations in research to actual product development. Nevertheless, important gaps between public and private support for different stages of technological development persist.

Overview of Selected Clean Energy Technology Options

This Appendix provides an overview[7] of the most promising technologies under development and in use today, including information on their strengths and weaknesses as well as their current stage of development (see table A2.1). It also provides an estimate of how much additional emission reduction could be realized from each technology versus a business-as-usual case. These estimates on potential emission reduction are drawn from the International Energy Agency publication *Energy Technology Perspectives* (IEA 2006a), which presents different scenarios of technological operation under accelerated technological development. The potential CO_2 reduction does not refer to the *total* amount a given technology can lower emissions but instead to the *additional* emission reduction per year in 2050 (compared with business-as-usual) that can be realized through the following actions:

- Increased support for research and development (R&D) for technologies with technical challenges and the need to reduce costs before commercial viability;
- Demonstration programs for energy technologies;
- Deployment programs for technologies not yet cost-competitive but whose costs could be reduced through learning-by-doing;
- Introduction of policies and measures that would attribute a price of $25 to each ton of CO_2 emitted into the atmosphere;
- Policies to overcome noneconomic barriers to technology commercialization such as standards, labeling information campaigns, and energy auditing.

---

7. Much of this overview of clean energy technologies comes from IEA 2006a and 2007b, REN21 2005 and Smil 2005 and World Bank staff estimates.

Table A2.1. Clean Energy Technologies and Mitigation Potential Resulting from Accelerated Technology Innovation

Technology	Stage of Development	Mitigation Potential by 2050 (Gt CO_2/year)	Comments
Supercritical, ultra-supercritical	R&D–Commercial	0.3	– Supercritical is commercial; ultra-supercritical requires more development, especially in the area of high-temperature materials
IGCC	R&D–Demonstration	0.2	– Enabling technologies for CCS
Carbon capture & storage	Demonstration	5.5	– Enabling technology for CCS – Cost barriers – Needs successful demonstrations of full system integration – Challenges for regulatory and legal systems
Hydropower	Scale-up–Commercial	0.5	– Large-scale is commercial – Mini and micro are demonstration/scale-up
Solar	R&D–Commercial	0.5	– PV is commercial in certain off-grid applications – Grid applications are in R&D phase, large cost reductions required – Concentrating solar power (CSP) is in demonstration phase
Ocean energy	R&D	0.1	– Early stages of development
Geothermal	Commercial	0.3	– Large potential in certain regions if costs can be reduced
Wind	Scale-up	1.3	– Deployment policies have significantly reduced costs but is still rarely commercial

Technology	Stage		Barriers
Bioelectricity	Commercial	0.5	– Large potential for BIG/GT, IGCC, and biorefineries but they are in R&D/demonstration stage
Hydrogen fuel cells for transport	R&D	0.8	– Very significant cost barriers
Second-generation biofuels	R&D–Demonstration	1.3	– Significant cost barriers
End-use energy efficiency			– The primary barriers facing end-use efficiency technologies relate to market barriers, inadequate regulations, capital constraints, and lack of information
Vehicles (engine, nonengine, and hybrid technologies)	Scale-up–Commercial	5.4	
Heating and cooling	Commercial	2.6	– Where these nontechnical barriers can be overcome, private industry is normally ready to conduct technical work to bring the bulk of products to commercialization
Electrical end-use and other	Scale-up–Commercial	4.6	
Other	Scale-up–Commercial	1.8	
Nuclear power generation			– Barriers of public acceptance, and political, regulatory, environmental, safety and financial issues of reactor safety, waste disposal, and nuclear proliferation
II and III generation	Commercial	1.8	
IV generation	R&D	1.9	– Large cost barriers

Source: Figures adapted from IEA 2006a.

Note: Figures indicate *additional* abatement potential of each technology relative to the baseline scenario, not total potential for CO_2 emission reduction. This potential can be realized through increased energy R&D, more extensive demonstration and deployment programs, and a set of policies that lead to adoption of technologies that reduce CO_2 emissions at $25/ton.

Efficient Coal-Fired Electricity Generation

Coal is the most prevalent and least expensive fossil fuel available. The estimated global coal reserve to production ratio is more than 200, implying that coal can continue to be consumed at current rates for more than two centuries, much longer than oil or gas. Many of the countries with the largest and fastest growing energy demand have substantial coal reserves, such as China and India. As a result, virtually all future energy scenarios predict the increased use of coal for power generation, notably in developing countries.

In this light, technologies that increase the efficiency of power generation from coal and/or facilitate capturing CO_2 emissions are of critical importance for reducing the carbon footprint of fossil fuels.

Supercritical and Ultra-Supercritical Steam Cycle Coal Plants

Supercritical and ultra-supercritical coal plants are defined by their steam temperatures. Supercritical plants use steam temperatures of 540°C and above, ultra-supercritical plants use 580°C and above. Supercritical plants can achieve overall efficiencies of up to 46 percent. Ultra-supercritical plants are currently achieving efficiencies up to about 50 percent and possibly about 55 percent through accelerated technology development in the future.

Integrated Gasification Combined Cycle (IGCC)

IGCC systems, which can process a variety of different feedstocks including coal, petroleum coke, biomass, and municipal solid waste, are receiving considerable public attention as a clean and efficient power-generation technology. By converting solid fuel to a gas and burning that gas in a combined-cycle plant, IGCC offers considerable improvements in electricity-generation efficiency over using the solid fuel in a simple steam loop. Current demonstration plants have efficiencies of about 45 percent, but efficiencies of 50 percent or higher are expected in the coming decades. In addition, emissions of a wide range of pollutants can be reduced by capturing them on-site.

Carbon Capture and Storage (CCS)

Given the continued importance of coal and natural gas for power plants, technologies that can reduce carbon emissions from electricity generation using fossil fuels is vitally important. The higher efficiency technologies described above could improve efficiencies by up to 10 percent or possible slightly more. However, although very helpful, such innovation is not enough to counteract emission increases due to expected rapid growth rates in fossil fuel power generation. This is where CCS holds such promise. If fully developed, it could reduce emissions from coal plants as much as 90 percent and thus allow continued and expanded use of coal within the framework of substantial global CO_2 emission reduction.

Carbon capture and storage reduces carbon emissions from fossil fuel combustion by separating out CO_2 from the combustion process and sequestering it in a permanent storage site so that it is prevented from reaching the atmosphere and contributing to global warming. The technology requires three distinct stages: (i) capturing CO_2 from power plants (or other concentrated CO_2 sources such as in the chemical industry), (ii) transporting captured CO_2 by pipeline or tanker, and (iii) storing CO_2 underground in deep saline aquifers, depleted oil and gas reservoirs, or coal seams. The technologies in each stage face different challenges; all these challenges must be addressed if technology is to be deployed widespread.

Renewable Power Generation

Renewable energy technologies offer an important option for generating electricity at little or no greenhouse gas emissions. A variety of technologies at different stages of development can reduce emissions. Many of these technologies require site-specific operating conditions (such as solar, ocean energy). Consequently, the focus of R&D efforts varies considerably between countries, reflecting domestic relevance of the various technologies.

Small and Large Hydropower

Hydropower is the dominant source of renewable electricity today and accounts for 19 percent of total electricity production and 82 percent of total renewable electricity (RENI21 2005). About 808 GW of hydropower capacity is currently in operation or under construction, the majority in Brazil, Canada, China, India, Norway, the Russian Federation, and the United States (IEA 2006a). In addition to electricity generation, hydropower projects are often designed to enhance potable and irrigation water supply and provide flood control. The majority of hydroelectricity is produced from large hydro projects but smaller hydro— most of which are run-off-river design—are increasingly being built, especially in China.

Solar Photovoltaic (PV) and Solar Thermal Electric

Solar photovoltaic electricity is a modular technology, based on semiconductors that convert sunlight directly into electricity. Each module usually produces up to several hundred watts but can be combined into large power arrays to fit different applications, grid-connected and off-grid. Currently, the globally installed capacity is about 4GW—most of which is in Germany, Japan, and the United States (RENI21 2005).

Solar thermal electric, also termed concentrating solar power (CSP), uses direct sunlight, which is concentrated through reflectors and used for heating and cooling applications or directly converted to electricity. In contrast to PV, CSP technologies can be integrated with conventional thermal cycles. The technology is, however, very demanding in terms of sunlight requirements and is therefore mostly relevant in arid, sun-intensive regions. Currently, about 400MW of CSP are in operation but several large-scale projects are being developed in Algeria, the Arab Republic of Egypt, Spain, and the United States (ibid.).

Ocean Energy

Different ocean energy technologies are being developed along four major concepts. Wave energy facilities harness kinetic energy associated with ocean waves. Tidal and marine current systems capture the potential energy associated with ocean currents and tides. Ocean thermal energy conversion (OTEC) extracts power by exploiting temperature differences between surface water and deep water. Salinity gradient (osmotic energy) systems harness the entropy of mixing freshwater and salt water, for example at river mouths.

Geothermal Power

Geothermal power plants harness geothermal heat to generate electricity. Different forms of geothermal technology are being developed, which differ in terms of the type of geothermal heat exploited (that is steam, hot water, or dry rock) and in their conversion technology. Most large-scale geothermal power development is currently limited to regions near tectonic plate boundaries but some technologies allow harnessing less accessible resources, for example, through deep drilling. Currently there are about 8GW of installed capacity worldwide (ibid).

Wind

Harnessing wind energy through on-shore and off-shore wind farms has been one of the fastest growing forms of renewable energy in the past decade—largely propelled by deployment subsidy policies in several OECD countries. Currently installed capacity is in the range of 48GW and second only to hydropower as a source of renewable electricity (ibid.).

Bioelectricity

Biomass fuels have similar characteristics to coal so that similar technologies are used to generate electricity. In this way, most biomass is combusted and used in steam cycles, either alone or co-fired with coal. Current global capacity is about 40GW, mostly installed in conjunction with agricultural processing (such as the sugar industry), forestry (such as paper mills), or landfill sites (ibid.).

Second Generation Biofuels

Biofuels refer to the production of liquid fuels from agricultural products. The fuels produced are mostly used as substitutes for petroleum products, primarily gasoline and diesel fuel. The current generation of biofuels is limited by the agricultural feedstocks that can be used (such as surgarcane, corn, and palm oil) and by potential negative social and environmental effects associated with their large-scale production. Moreover, their production is economically viable only in the most efficient feedstock-producing countries and under particularly high oil prices, such as sugar cane ethanol in Brazil. Second-generation biofuels have promise because they would be able to convert any biomass material into liquid fuels, generate high emissions reductions, and mitigate competition between biofuel and food production.

Second-generation biofuel technologies such as lingocellulosic ethanol and biomass-to-liquids (BTL) biodiesel allow the conversion into biofuels not only of the glucose and oils retrievable in today's first-generation bioethanol and biodiesel, but also of cellulose, hemicellulose, and even lignin—the main building blocks of most biomass. Thus, less expensive and more abundant feedstocks such as residues, stems, and leaves of crops, straw, biodegradable urban wastes, weeds, and fast-growing trees can in principal be converted into biofuels. This considerably increases the achievable scale of production and mitigates many of the social and environmental concerns associated with first-generation biofuels.

Energy Efficiency

End-use energy efficiency comprises a broad set of technologies and technology management systems that can achieve the same energy services with lower energy inputs. Most attention is currently drawn to improvements in efficiency in the transport (mostly automobiles) and buildings sectors as well as in the industry sector. For automobiles, for example, research goals include reducing vehicle weight, improving aerodynamics, and improving the efficiency of engine components. For buildings, efficiency measures include efficient lighting (such as CFLs and LEDs); better insulation; and more efficient heating, cooling, and water pumping systems. Most industrial processes can be optimized with regards to their conversion efficiency of energy inputs, such as through more efficient motor systems.

Nuclear Fission

Nuclear power provides a significant amount of electricity in France and in numerous countries, including China, Finland, India, Japan, Russia, Sweden, the United Kingdom, and the United States. More than 430 nuclear power plants are operating in the world, accounting for 17 percent of the world's electricity generation in 2002. In OECD countries, 346 reactors were connected to the grid at the end of 2006, constituting 23.1 percent of the total OECD and 30 percent of the European Union electricity supply. With climate change issues growing in importance, a new debate over an increasing role of nuclear power has been launched in several countries. Nuclear power can offer a positive contribution to energy security because most reserves for the uranium and thorium used in nuclear technologies are not located in sensitive regions. In the short to medium term, the lifetime of existing plants could be extended from the initial 40 years to up to 60, depending on the type and use of the power plant. However, nuclear waste disposal, reactor safety, and nuclear proliferation as well as related liability issues continue to be of considerable concern.

Hydrogen Fuel Cells for Transport

Fuel cells powered by hydrogen have received considerable public attention, especially for powering cars. Both for stationary and automotive applications, fuel cells are significantly more expensive than competing technologies. Several demonstration vehicles have been developed, but the technology is still in a relatively early R&D phase. Although in the automotive sector the high efficiency makes fuel cells equal or less expensive than gasoline operations, the capital costs of fuel cells are considerably higher than for gasoline-powered vehicles. Major technological advances and cost reductions are necessary in all areas. Research is focusing especially on the development of membranes, hydrogen storage components, and improvements in stack life. Once these barriers are overcome and it becomes clear what kind of on-board hydrogen storage technology will prevail, additional R&D efforts are needed on hydrogen refueling infrastructure.

Analyses Supporting the Need for Technological Innovation

IPCC Fourth Assessment Report

The Intergovernmental Panel on Climate Change Working Group 3 report on climate change mitigation released in May (IPCC 2007a) strongly supports the need for new and improved energy technologies to achieve a sustainable future. This report states that "investments in and worldwide deployment of low-GHG emission technologies *as well as* technology improvements through public and private research, development, and deployment (RD&D) would be required for achieving stabilization targets" (italics added).

The report also lists the key mitigation technologies that need to be commercialized by 2030 to stabilize emissions, which include the following:

- Carbon capture and storage (CCS) for gas, biomass, and coal-fired generating facilities;
- New and improved forms of renewable energy, including tidal and waves energy, concentrating solar, and solar PV;
- Improved energy efficiency;
- Second-generation biofuels;
- Higher efficiency aircraft;
- Advanced electric and hybrid vehicles;
- CCS for cement, ammonia, and iron manufacture; and
- Advancement in agricultural technologies.

None of these technologies is currently commercially available to deploy on any type of significant scale. Getting them ready for substantial deployment would require a concerted RD&D effort.

The Stern Review

The Stern Review (HM Treasury 2006) also highlights the need for new and improved energy technologies to address global warming. The Review notes that stabilization of atmospheric greenhouse gas concentrations—at whatever level—will require reducing annual global emissions below 5 $GtCO_2e$, which is the level that the earth can naturally absorb without adding to atmospheric concentrations. This level is more than 80 percent below the absolute level of current annual emissions. To achieve such reductions, the Review states that policies to accelerate new technology development will be one of three essential elements to combat climate change.[8]

More specifically, the Stern Review states that tackling climate change "requires a widespread shift to new or improved technology in key sectors such as power generation, transport, and energy use" and that "the development and deployment of a wide range of low-carbon technologies is essential in achieving the deep cuts in emissions that are needed." The Review explains that although the private sector plays the major role in R&D and technology diffusion, enhanced collaboration between government and industry will further stimulate the development of a broad portfolio of low-carbon technologies and reduce costs. It calls for a doubling of global public energy R&D funding, to about $20 billion per year, and an increase of two to five times globally for deployment incentives for low-emission technologies.

IEA Scenarios

International Energy Agency scenarios demonstrate the limitations of policies that rely solely on current technologies. In its alternative policy cenario (APS), IEA assumes that all policies currently under consideration around the world to reduce emissions are fully implemented (IEA 2006b). This scenario assumes continuing improvements in the cost and performance of energy technologies, but does not include any efforts to accelerate these improvements beyond current trends. Consequently, two promising technologies—carbon capture and storage and second-generation biofuels—are not assumed to be commercially available by the end of the 2030 scenario window.

The APS—with its mix of extensive policy implementation but modest technological advances—does not present a sustainable energy future. From 2004 to 2030, global energy demand rises by 38 percent and CO_2 emissions by 31 percent. By 2030, oil still accounts for 32 percent of total supply while the combined supply of all fossil fuels is 77 percent of the total. These figures—though improvements from business-as-usual due to the extensive policy implementations—underline the limitations of relying on today's energy technologies.

The IEA has also examined scenarios that assume accelerated development of clean energy technologies, found in *Energy Technology Perspectives* (IEA 2006a). The IEA uses accelerated technology scenarios (ACTs) where increased R&D and other measures bring clean energy technologies to commercialization more rapidly than would be the case under business-as-usual. The advanced technologies include renewable energy, end-use efficiency

---

8. The other two elements are carbon pricing and the removal of barriers to behavioral change.

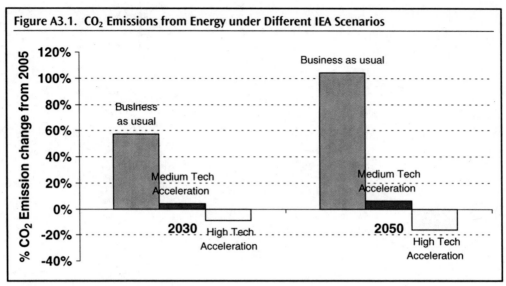

Figure A3.1. CO_2 Emissions from Energy under Different IEA Scenarios

Source: Adapted from IEA 2006a.

Note: The mitigation policies in the third scenario are generally stronger than in the second scenario. The lower emission in the third scenario would result from both more stringent polices and accelerated energy technology development.

technologies, carbon capture and storage (CCS), nuclear power, and improved efficiency of fossil fuel-fired generating stations. The ACT projections show that accelerated technology development can make a substantial difference in our energy future. Under the accelerated technology scenario, CO_2 emissions rise just 6 percent from 2003 to 2050. Although this scenario also assumes policies to encourage cleaner energy use (such as a $25/ton value put on CO_2 emissions), the major difference between it and the other less-promising scenarios is the added push for energy technology innovation. Figure A3.1 shows emission scenarios for these three projections.

Historical Data on Government Energy R&D Spending

This appendix provides a more detailed picture of public and private R&D spending in IEA countries. Tables A4.1 and A4.2 depict the technologies where R&D has been focused over the past 25 years. Figures A4.1 and A4.2 show how—until recently—energy R&D has been falling both in absolute terms and relative to other industries.

Summary of Recent Energy RD&D Activities of Japanese and U.S. Governments

As depicted in figure A4.3, Japan and the United States are by far the largest contributors to energy R&D in absolute terms. Thus, the following section briefly outlines some of the most recent RD&D activities in these two countries.

The United States, as a result of its Advanced Energy Initiative (AEI) and America COMPETES Act, increased both basic energy science (by 5 percent) and applied energy R&D (by 32 percent) in FY07. These increases included dramatic boosts for selected renewable energy priorities, including hydrogen, solar power, and biomass. Increases in these priorities are set to continue in FY08; the U.S. Congress also proposes to increase energy R&D spending dramatically for other renewable energy, fossil fuels, and energy conservation programs. Additionally, $300 million has been authorized in 2008 for the new Advanced Research Projects Agency for Energy (ARPA-E) to fund breakthrough alternative energy R&D technologies. Overall, U.S. energy R&D spending rose 11 percent (in real terms) in FY07 with a projected increase of 18 percent to in FY08.

In addition, in September 2007, the U.S. government proposed the creation of a new international clean technology fund to help developing nations harness the power of clean

Table A4.1. Public R&D Expenditures in IEA Countries

	R&D Budgets (million 2005 US$, exchange rates)				Share of Total R&D (%)	
	1992–2001	2002	2003	2004[a]	1992–2001	2004[a]
Energy efficiency	12,206	1,627	1,215	1,094	13.24%	11.61%
Fossil fuels	9,545	1,099	1,010	1,089	10.35%	11.56%
Oil & gas	4,375	603	525	500	4.74%	5.31%
Coal	5,170	495	484	520	5.61%	5.52%
CO_2 capture and storage	—	—	—	69	0.00%	0.73%
Renewable energy sources	7,683	896	868	1,082	8.33%	11.48%
Solar energy	3,865	439	417	502	4.19%	5.33%
Wind energy	1,162	118	123	128	1.26%	1.36%
Ocean energy	63	5	4	9	0.07%	0.10%
Bio-energy	1,657	229	236	296	1.80%	3.15%
Geothermal energy	771	76	58	47	0.84%	0.50%
Hydropower	165	27	28	31	0.18%	0.33%
Other renewables	—	—	—	66	—	0.71%
Nuclear fission and fusion	44,939	4,204	4,222	3,763	48.73%	39.91%
Hydrogen and fuel cells	—	—	—	253	0.00%	2.69%
Other power & storage	4,181	523	506	373	4.53%	3.96%
Other[b]	13,662	1,815	1,875	1,772	14.81%	18.80%
Total energy R&D	92,217	10,165	9,699	9,429	100.00%	100.00%

Source: IEA databases.

Note: — = Not available.

a. 2004 is the last year for which detailed and updated R&D statistics for all major IEA countries was available in IEA databases at the time of writing.

b. "Other" includes energy systems analysis (system analysis related to energy R&D; sociological, economical, and environmental impact of energy not specifically related to a technology listed above) as well as hydrogen, energy technology information dissemination, and studies not related to a specific technology area listed above.

energy technologies. This fund will help finance clean energy projects in the developing world. The U.S. government began it by the end of 2007.

Several Japanese initiatives also point to renewed commitment to clean energy technologies. The "New National Energy Strategy" released by the Ministry of Economy, Trade, and Industry (METI) in 2006 aims to establish the foundation for sustainable development through a comprehensive approach for energy issues and environmental issues. Targets include a 30 percent increase in energy efficiency by 2030 through a positive cycle of technological innovation and new energy innovation through the development of revolutionary energy technologies including cleaner-burning coal and new-generation solar technology. The recently announced "Cool Earth 50" initiative, aimed at creating a new international framework to fight global warming beyond the 2012 expiration of the Kyoto Protocol, includes investments in clean energy technologies. Japan's "Innovation 25" program sees a clean energy industry based on new technologies as a source of economic growth.

Table A4.2. Combined Spending by All IEA Governments on Energy R&D (million of US$ in 2004 prices and exchange rates)

Energy Category	1974	1975	1976	1977	1978	1979	1980	1981	1982	1983	1984
Conservation/efficiency	167	289	372	608	729	756	1,016	837	728	915	848
Fossil fuels	405	569	852	1,326	1,629	1,765	2,551	2,613	1,634	1,600	1,506
Renewable energy	68	216	370	805	1,229	1,724	2,029	2,007	1,222	1,094	1,069
Nuclear fission	3,877	4,497	4,340	5,794	6,133	6,559	6,573	6,591	6,363	6,086	5,809
Nuclear fusion	435	577	734	1,059	1,309	1,577	1,267	1,368	1,445	1,423	1,433
Power & storage	141	172	199	391	502	716	430	355	275	318	296
Other tech./research	829	920	1,214	910	1,104	1,006	1,276	461	422	896	848
Total	**5,922**	**7,239**	**8,080**	**10,894**	**12,636**	**14,103**	**15,142**	**14,233**	**12,090**	**12,332**	**11,808**

Energy Category	1985	1986	1987	1988	1989	1990	1991	1992	1993	1994	1995
Conservation/efficiency	861	735	773	641	569	626	750	694	818	1,109	1,217
Fossil fuels	1,513	1,520	1,350	1,494	1,375	1,851	1,686	1,171	1,202	1,229	1,035
Renewable energy	893	711	624	618	560	565	747	689	717	746	800
Nuclear fission	6,807	6,345	5,030	4,294	4,783	4,274	4,476	3,754	3,641	3,526	3,657
Nuclear fusion	1,507	1,355	1,275	1,183	1,094	1,066	1,103	1,056	1,158	1,115	1,110
Power & storage	300	274	290	341	360	283	318	269	270	385	366
Other tech./research	895	790	949	1,118	1,262	1,119	1,350	1,227	1,383	1,254	1,241
Total	**12,776**	**11,729**	**10,290**	**9,689**	**10,003**	**9,785**	**10,429**	**8,860**	**9,189**	**9,364**	**9,426**

(continued)

Table A4.2. Combined Spending by All IEA Governments on Energy R&D (million of US$ in 2004 prices and exchange rates) (*Continued*)

Energy Category	1996	1997	1998	1999	2000	2001	2002	2003	1974–2003 Total
Conservation/efficiency	1,143	1,100	1,304	1,400	1,471	1,603	1,621	1,189	26,889
Fossil fuels	966	818	647	665	515	734	1,021	971	38,214
Renewable energy	717	705	775	761	751	823	888	841	25,767
Nuclear fission	3,601	3,457	3,370	3,372	3,404	3,205	3,539	3,199	140,352
Nuclear fusion	999	965	896	723	877	823	691	655	32,279
Power & storage	351	375	419	399	549	622	524	510	10,997
Other tech./research	1,224	1,275	1,211	1,320	1,271	1,575	1,767	1,836	33,952
Total	**9,000**	**8,696**	**8,622**	**8,641**	**8,838**	**9,385**	**10,051**	**9,200**	**308,451**

Source: IEA Online R&D Statistics Database, http://www.iea.org/rdd/default.aspx; and correspondence with Richard Doornbosch, OECD.

Figure A4.1. Public R&D Spending in OECD Countries (US$ billion)

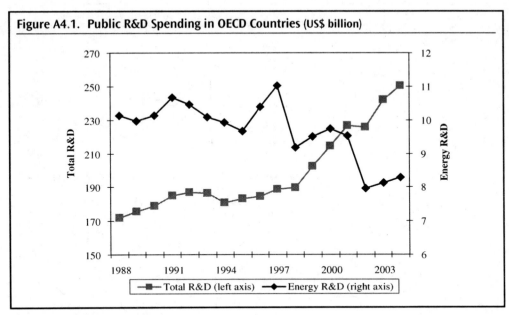

Source: Doornbosch and Upton 2006.

Figure A4.2. Total U.S. Public Sector Energy R&D Investment, 1974–2006

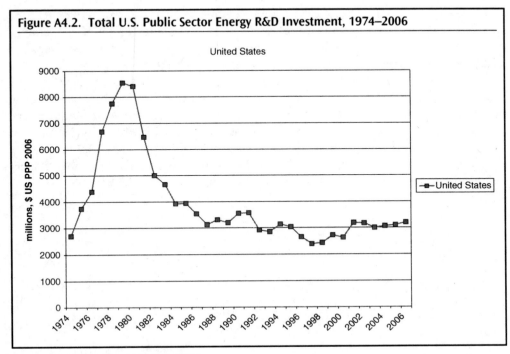

Source: Doornbosch and Upton 2006.

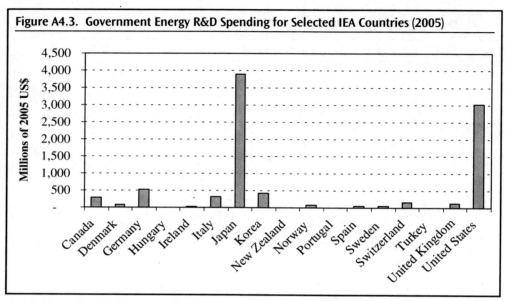

Figure A4.3. Government Energy R&D Spending for Selected IEA Countries (2005)

Source: IEA online R&D Statistics Database, http://www.iea.org/rdd/default.aspx.

There have also been substantial increases in other means of government support for clean energy technologies beyond those associated with traditional R&D programs. Primary among these are support systems for renewable energy that can take the form of direct subsidies, feed-in tariffs, renewable portfolio standards, tax rebates, and biofuel blending mandates. These support policies may at least in part compensate for the relatively low R&D figures described above and have been increasing almost without exception in OECD countries in the last two to three years. It is estimated that such deployment incentives amount to $33 billion globally (HM Treasury 2006). Such support leads to fuller deployment of technologies not yet commercially competitive, such as wind power, and in this way brings manufacturing and operating experience that leads to more reliable, lower-cost technology. The increase in such support systems thus acts as a *de facto* technology acceleration program although it must be noted that—in contrast to RD&D programs—they apply only to technologies with near-term commercial viability and reliability.

Bibliography

Barton, John H. 2007. "New Trends in Technology Transfer." International Centre for Trade and Sustainable Development Intellectual Property and Sustainable Development Series, Issue Paper 18.

Brown, John S. and John Hagel III. 2006. "Creation Nets: Getting the Most from Open Innovation." *McKinsey Quarterly* 2:40–51.

Carr, Nicholas G. 2007. "The Ignorance of Crowds." *Strategy + Business.* Summer.

CGIAR (Consultative Group on International Agricultural Research). 2006. *Annual Report 2005. Science* Based *Solutions: The Science Behind Growth and Development.* Washington, D.C.

———. 2007. *Annual Report 2006. Focus on Partnerships for Effective Research.* Washington, DC.

Collins, F. S., M. Morgan, and A. Patrinos. 2003. "The Human Genome Project: Lessons from Large-Scale Biology." *Science* 300:286–90.

DOE (U.S. Department of Energy). 2001. "Genomics and Its Impact on Science and Society; The Human Genome Project and Beyond." http://genomics.energy.gov.

———. 2006. "Human Genome Project Information." www.ornl.gov/hgmis.

Dooley, J. J., and P. J. Runci. 2004. "Energy and Carbon Management R&D: Key Considerations and Recommendations for Policy." Commissioned paper PNWD3435 for the National Commission on Energy Policy, Joint Global Research Institute, Battelle Pacific Northwest Division. College Park, MD.

Dooley, J. J., P. J. Runci, and Leon Clarke. 2006. *Trends in Energy R&D.* Battelle, Joint Global Change Research Institute.

Doornbosch, R., and S. Upton. 2006. *Do We Have the Right R&D Priorities and Programmes to Support the Energy Technologies of the Future?* Paris: OECD.

EIA (Energy Information Agency). 2006. *International Energy Outlook 2006.* Washington, D.C.

Farlow, A. W. K., D. W. Light, Richard T. Mahoney, and Roy Widdus. 2005. "Concerns Regarding the Center for Global Development Report 'Making Markets for Vaccines.' " Submission to the Commission on Intellectual Property Rights, Innovation, and Public Health: WHO.

Finkelstein, A. 2004. "Static and Dynamic Effect of Health Policy: Evidence from the Vaccine Industry." *Quarterly Journal of Economics* 119:527–64.

Foxon, Timothy J. 2003. "Inducing Innovation for a Low-Carbon Future: Drivers, Barriers, and Policies." A report for the Carbon Trust. London.

Gagnon-Lebrun, F. 2004. "International Energy Technology Collaboration and Climate Change Mitigation. Case Study 2: Cooperation in Agriculture: R&D on High-Yielding Crop Varieties." Paris: OECD/IEA.

Gallagher, K. S., J. P. Holdren, and Ambuj D. Sagar. 2006. "Energy-Technology Innovation." *Annual Review of Environment and Resources* 31:193–237.

Garibaldi, J. A. 2007. *Scaling Up Responses to Climate Change. Technology and R&D Investment and an Environment for a Low Carbon Technology Deployment.* http://unfccc.int/files/cooperation_and_support/financial_mechanism/application/pdf/garibaldi.pdf.

Global Forum for Health Research. 2004. *10/90 Report on Health Research.* Geneva.

———. 2006. *Monitoring Financial Flows for Health Research 2006. The Changing Landscape of Health Research for Development.* Geneva.

Hippel, Eric von and Georg von Krogh. 2003. "Open Source Software and the "Private-Collective" Innovation Model: Issues for Organizational Science." *Organizational Science* 14(2):20–23.

HM Treasury. 2006. *The Economics of Climate Change, The Stern Review.* London.

Holdren, John P. 2006. "The Energy Innovation Imperative. Addressing Oil Dependence, Climate Change, and Other 21st Century Energy Challenges." *Innovations* (Spring): 313.

IEA (International Energy Agency). 2004. *Renewable Energy. Market and Policy Trends in IEA Countries.* Paris.

———. 2006a. *Energy Technology Perspectives. In Support of the G8 Plan of Action. Scenarios & Strategies to 2050.* Paris.

———. 2006b. *World Energy Outlook 2006.* Paris.

———. 2007a. *CO_2 Emissions form Fuel Combustion.* Paris

———. 2007b. *Energy Technologies at the Cutting Edge.* Paris.

———. 2007c. *World Energy Outlook 2007. China and India.* Insights. Paris.

Inter Academy Council. 2007. *Lighting the Way. Towards a Sustainable Energy Future.* Amsterdam.

IPCC (Intergovernmental Panel on Climate Change). 2007a. *Climate Change 2007: Impacts, Adaptation, and Vulnerability. Working Group II Contribution to the Intergovernmental Panel on Climate Change. Fourth Assessment Report.* New York: Cambridge University Press.

———. 2007b. "Summary for Policymakers." In *Climate Change 2007: Mitigation. Contribution of Working Group III to the Fourth Assessment Report of the Intergovernmental Panel on Climate Change,* ed. B. Metz, O. R. Davidson, P. R. Bosch, R. Dave, and L. A. Meyer, 3–23. New York: Cambridge University Press.

———. 2007c. *Working Group III Contribution to the IPCC Fourth Assessment Report, Summary for Policymakers.* New York: Cambridge University Press.

IPIECA (International Petroleum Industry Environmental Conservation Association). 2006. "Increasing the Pace of Technology Innovation and Application: Enabling Climate Change Solutions." http://www.ipieca.org/activities/climate_change/downloads/publications/PaceOfTechnology.pdf.

Karplus, Valerie J. 2007. "Innovation in China's Energy Sector." Working Paper 61, Program on Energy and Sustainable Development, Stanford University, Stanford, Calif.

Kogut, Bruce and Anca Metiu. 2001. "Open-Source Software Development and Distributed Innovation." *Oxford Review of Economic Policy* 17 (2):248–64.

Lambright, W. H. 2002. *Managing "Big Science": A Case Study of the Human Genome Project.* Available by The PricewaterhouseCoopers Endowment for the Business of Government, http://www.businessofgovernment.org/pdfs/LambrightReport2.pdf.

Levine, R., M. Kremer, and Alice Albright. 2005. "Making Markets for Vaccines. Ideas to Action." The report of the Center for Global Development Advanced Market Commitment Working Group. Washington, D.C.: CGD.

Light, D. W. 2005. "Making Practical Markets for Vaccines." *PLoS Medicine* 2 (10).

Milford, L Cohen A. and Barker, T. 2007. *Distributed Technology Innovation: A New Business Model for Climate Stabilization.* Montpelier, VT: Clean Energy Group.

NIH (National Institutes of Health). 2007. National Human Genome Research Institute. http://www.genome.gov.

Pacala, S., and R. Socolow. 2004. "Stabilization Wedges: Solving the Climate Problem for the Next 50 Years with Current Technologies." *Science* 305:968–72.

Pardey, P. G., N. Beintema, Steven Dehmer, and Stanley Wood. 2006. *Agricultural Research. A Growing Global Divide?* Washington, D.C.: IFPRI.

Quiggin, John. 2006. "Blogs, Wikis and Creative Innovation." *International Journal of Cultural Studies* 9(4):481–96.

REN21. 2005. *Renewables 2005 Global Status Report.* Washington, D.C.: Worldwatch Institute.

Roberts, Leslie, R. John Davenport, Elizabeth Pennisi, and Eliot Marshall. 2001. "A History of the Human Genome Project." *Science* 291 (5507):1195–201.

Smil, V. 2005. *Energy at the Crossroads. Global Perspectives and Uncertainties.* Cambridge, Mass.: MIT Press.

UNEP (United Nations Environment Programme). 2007. *Global Trends in Sustainable Energy Investment 2007: Analysis of the Trends and Issues of Renewable Energy and Energy Efficiency in OECD and Developing Countries.* Paris: UNEP Sustainable Energy Initiative (SEFI) and New Energy Finance Ltd.

Weber, Steven. 2000. "The Political Economy of Open Source Software." BRIE Working Paper 140. E-conomy Project Working Paper 15. http://repositories.cdlib.org/cgi/viewcontent.cgi?article=1011&context=brie.

Wikipedia. 2007. Articles on the keywords: Wiki, Creative Commons, Open Source, Network Innovation, Linux, Wikipedia, Open Design. http://www.wikipedia.org.

World Bank. 2003. *An Independent Meta-Evaluation of the Consultative Group on International Agricultural Research. Volume 1: Overview Report.* Operations Evaluation Department. Washington, D.C.

———. 2007a. *Clean Energy for Development Investment Framework: Progress Report on the World Bank Group Action Plan.* Washington, D.C.

————. 2007b. *International Trade and Climate Change. Economic, Legal, and Institutional Perspectives.* Washington, D.C.

————. 2008. *Global Economic Prospects 2008: Technology Diffusion in the Developing World.* Washington, D.C.

World Energy Council. 2007. *Energy and Climate Change. Promoting the Sustainable Supply and Use of Energy for the Greatest Benefit of All.* London.

Worldwatch Institute. 2007. *Biofuels for Transportation: Global Potential and Implications for Sustainable Agriculture and Energy in the 21st Century.* Washington, D.C.

Eco-Audit

Environmental Benefits Statement

The World Bank is committed to preserving Endangered Forests and natural resources. We print World Bank Working Papers and Country Studies on 100 percent postconsumer recycled paper, processed chlorine free. The World Bank has formally agreed to follow the recommended standards for paper usage set by Green Press Initiative—a nonprofit program supporting publishers in using fiber that is not sourced from Endangered Forests. For more information, visit www.greenpressinitiative.org.

In 2007, the printing of these books on recycled paper saved the following:

Trees*	Solid Waste	Water	Net Greenhouse Gases	Total Energy
264	12,419	96,126	23,289	184 mil.
˙40" in height and 6-8" in diameter	Pounds	Gallons	Pounds CO_2 Equivalent	BTUs